AF572505

# Elastic Wave Scattering and Propagation



**Edited by**

**Vijay K. Varadan**
**Vasundara V. Varadan**

230 Collingwood, P.O. Box 1425, Ann Arbor, Michigan 48106

Library of Congress Catalog Card Number 81-70871
ISBN 0-250-40534-2

Manufactured in the United States of America


Butterworths, Ltd., Borough Green, Sevenoaks
Kent TN15 8PH, England

# PREFACE

With advances in high-speed digital computers and innovations in experimental techniques, our understanding of problems related to elastic wave propagation and scattering has increased significantly. It is well known that these studies have important applications in many areas of physics and engineering such as geophysical exploration, ultrasonic nondestructive evaluation of flaws, ultrasonic characterization of inhomogeneous materials, medical ultrasonics and underwater acoustics.

At the Midwestern Mechanics Conference held at the University of Michigan, Ann Arbor, Michigan May 7–9, 1981, the editors of this volume organized a special session on elastic wave propagation and scattering. Scientists from government, industry and universities were invited to assess the state of the art and to outline their recent work. They were also generous enough with their time and contributed written versions summarizing the presentations made at Ann Arbor.

The topics covered include experimental techniques, analytical and numerical studies of scattering from single and finite numbers of obstacles as well as statistical theories of wave propagation in inhomogeneous media. We hope that this volume will be a useful reference to every researcher in the field of elastic wave propagation.

The editors wish to thank the contributors to this book for both their oral and written presentations. We are indebted to Professors R. A. Scott and A. S. Wineman of the University of Michigan for their encouragement in the planning of the symposium that has resulted in this book. We also wish to thank Dr. P. Chandran and Dr. T. A. K. Pillai for their assistance with the figures and references, and our typist, Toni DiGeronimo, for her patience and cooperation. Thanks are due to Ann Arbor Science Publishers for publishing this book.

Vijay K. Varadan
Vasundara V. Varadan

**Vijay K. Varadan** is Associate Professor, Department of Engineering Mechanics, Ohio State University. Dr. Varadan received his PhD in theoretical and applied mechanics from Northwestern University, his MS in engineering mechanics from Pennsylvania State University, and his BS in mechanical engineering–machine design from the University of Madras, Madras, India. He is a member of the American Society of Mechanical Engineers and the Acoustical Society of America, and is a consultant to the Physical Acoustics Division at the Naval Research Laboratory, Washington, DC, through the Center of Excellence. Dr. Varadan is co-editor of *Acoustic, Electromagnetic and Elastic Wave Scattering—Focus on the T-Matrix Approach*, is co-author of numerous journal articles and conference proceedings, and serves as referee to a number of national and international journals. His research interests include: wave propagation, scattering, non-destructive testing and ultrasonics; seismology; fracture mechanics; dynamic crack propagation, branching and arrest; and underwater acoustics.

**Vasundara V. Varadan** is Associate Professor, Department of Engineering Mechanics, Ohio State University. Dr. Varadan's research interests include: wave propagation and scattering, numerical methods, applications, and unified approaches to classical field problems, both analytical and numerical. She has developed the T-matrix or Null Field Method to solve scattering problems by single, multiple and statistical ensembles of scatterers. Dr. Varadan is co-author of several journal articles, conference proceedings and abstracts, and has been invited to speak at various lectures, seminars and workshops. Together with Dr. Vijay K. Varadan, she organized an international symposium, "Recent Developments in Classical Wave Scattering—Focus on the T-matrix Approach" and co-edited the proceedings. Dr. Varadan is a consultant to the Physical Acoustics Division at the Naval Research Laboratory, Washington, DC, through the Center of Excellence and is a member of the Acoustical Society of America, Society of Engineering Science and Sigma Delta Epsilon. She received her PhD and MS in physics from the University of Illinois, and her MSc in solid state physics and BSc in physics and mathematics from the University of Kerala, Cochin, India.

# CONTENTS

ELASTIC WAVE SCATTERING AND PROPAGATION

CHAPTER 1

# UNDERWATER ACOUSTIC REFLECTION BY BODIES AND SURFACES: WHAT HAS BEEN DONE, IS BEING DONE, AND WHAT NEEDS DOING -- ONE MAN'S VIEW

*Werner G. Neubauer*
Naval Research Laboratory
Code 5108
4555 Overlook Avenue, S. W.
Washington, D. C. 20375

## ABSTRACT

The Navy uses reflection and scattering of elastic waves in ways that span a broad range of technical areas. They range from the ultrasonic noninvasive visualization of parts of the human body and detection of material flaws in mechanical parts, to the detection of submarines in the ocean. Also, other technologies such as electromagnetic reflection and scattering have similarities and differences from the analogous elastic problems, some of which will be discussed. In the past, the status of technology and understanding applicable to elastic reflection problems has been summarized and appraised and directions in which future effort would profitably proceed have been spelled out. How that matched what was done in the ensuing years will be discussed, as well as the status of present-day efforts in the same context. A subjective appraisal will be attempted of basic and applied research task areas worth consideration, regardless of whether they are presently doable; and those that seem in need of doing for the Navy's benefit will be pointed out.

## INTRODUCTION

Dependence of the Navy on acoustics to communicate, navigate and find objects in the ocean has given it a longstanding interest in reflection of scalar waves. Existing and planned Navy systems depend on the predictability of reflected wave fields in water, as well as the vibrational behavior of bodies. Although large strides were made with the pressure of WW II, the insight into acoustic reflection was very rudimentary, especially for problems that involved the elastic response of bodies and surfaces.

In 1948 a comprehensive appraisal of what was done during the war and where it was likely to lead was carried out and described in a now declassified document [1]. In that report L. Fussel, Jr. stated in an article that indicated great insight as well as foresight that "...it is felt that, on the whole, the aim of future research should be to (a) build up a backlog of reliable measurements of echoes under simple controlled conditions, (b) build up a unified theory of these simple effects and tie the experimental results into a coherent whole, and (c) learn to use these experimental and theoretical results in predicting the outcome of complex situations where control is difficult. The end purpose of research on echoes is to understand and predict echo ranging in the ocean..." Promising areas of study were described by Fussel, and at one point he says: "The solution to the problem must not be expected to come directly. It is more likely to grow through the consideration of a great many special cases, which will gradually point the way toward these properties of an echo which are significant in determining the properties of the target which produce it." These and other such statements pretty clearly foretold exactly what was to happen and has happened. In addition to analytical computation, experimental measurements in the laboratory went along hand-in-hand. In general, the largest variety of geometric shapes can be treated exactly or approximately if they are considered as reflection problems in which the body has soft or rigid boundary conditions, that is, the sound does not penetrate into the body. This is indicated in Figure 1.

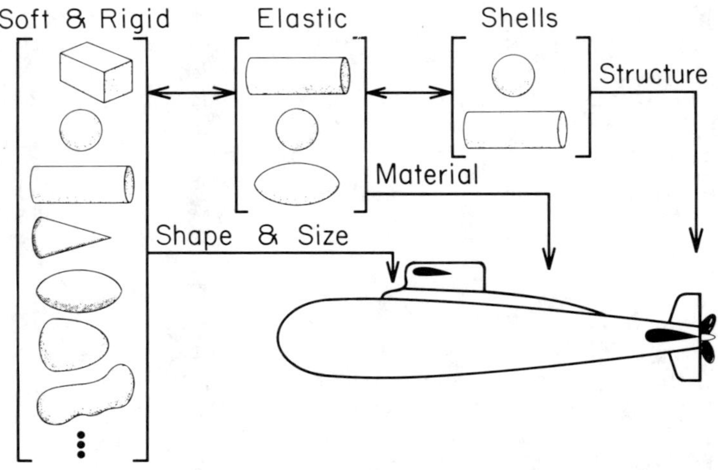


Figure 1. Reflection of sound by finite bodies.

However, when the material of the body is considered as well, and energy is allowed to penetrate into the body, the number of elastic solutions that can be dealt with is significantly smaller. Even approximate solutions available for shape alone fall by the wayside.

Generally, the stated end goal of work in reflection has been to achieve the means of computing reflection from a totally arbitrary shape; and consistent with the earlier rigid boundary conditions for the formulations in WW II, generalization to an arbitrary shape was attempted by computations using the Kirchhoff approximation in the absence of elasticity. Such approximation leaves a great deal to be desired, as can be seen on Figure 2, where Kirchhoff approximation and the exact theory for a rigid sphere are shown. Although many

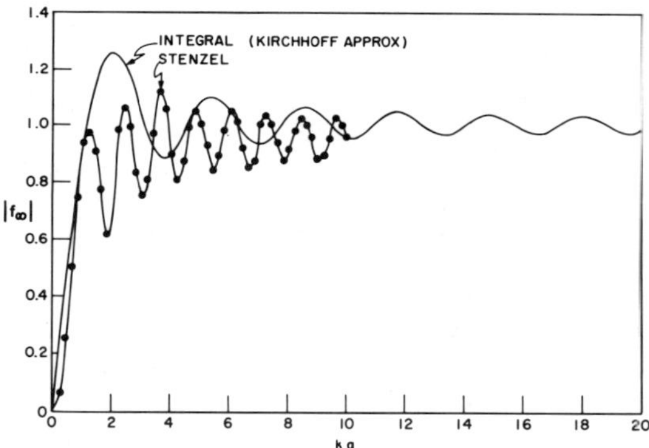


Figure 2. The reflection from a rigid sphere. (From W. G. Neubauer, J. Acoust. Soc. Am. 35, 279-285 (1963), with permission of the American Institute of Physics.)

titles in the literature proclaim the achievement of the goal of computations of reflection from an arbitrary shape, such solutions are certainly not available for elastic boundary conditions. If one goes to a shell, so that the structure of the shape is included , means of solution become even fewer. All three factors of shape and size, material, and structure are important in the reflection from an object of interest to the Navy, such as a submarine. Of course, specific problems can be treated by finite element methods, and others and recent success with T-matrix methods have shown real promise.

I will briefly try to trace the progress that has been made since 1948 by using spheres and cylinders as an example of general capabilities.

Limited harmonic series calculations were available for spheres, spheroids, and cylinders. Computational capability of reflection without the aid of computers was an extreme chore in 1951 [2], even for arguments of special functions, such that the reflector size is less than the acoustic wavelength. As computer capabilities increased, by 1958 [3], the problem of sphere reflection was carried out and computations for the first time exceeded understanding. That is, many more computations were available whose physical interpretation was not completely understood. Even so, computations were still severely limited by the lack of ability to compute special functions. In this case, spherical Bessel functions of arguments above 30 where the sphere radius would be less than 5 incident wavelengths large.

To give some orientation about Navy interests, at a frequency of 24 kHz, a submarine sized object of say 250 ft length would require function computations for arguments of over 400.

It was not until 1978 when the special function argument magnitude limitation essentially disappeared and reflection from elastic spheres and cylinders up to ka's of 950 were accomplished by Flax [4], where k is the incident acoustic wave number and a is the sphere radius. Those results showed that the solutions continue to oscillate significantly, even when the elastic body is many many wavelengths large. Before that time, it was assumed that the body behaved as indicated by the asymptotic solution to approach a more or less steady value. In any case, now we really consider any spherical, and infinite cylinder problem as well, one that can be done. This also includes multilayered shapes as well as layers that have losses in them.

## SCHLIEREN VISUALIZATION OF REFLECTION

In the following it would be helpful to utilize schlieren photographs to demonstrate acoustic fields. For that purpose I will describe just briefly how schlieren waves permit viewing of acoustic fields. In Figure 3 a schlieren system is outlined with a lens on either side of a water tank. A parallel

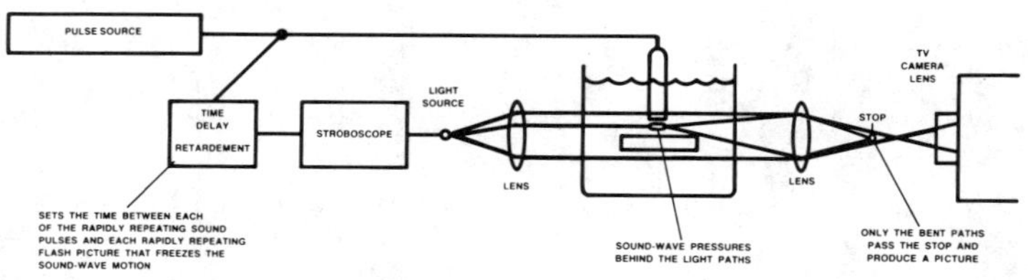


Figure 3. A stroboscopic schlieren system.

light beam is generated between these two lenses from a light source on the left which is very carefully focused to a stop at the focus of the lens on the right. In the absence of a sound wave in the water one thus achieves a dark field. When a sound wave is introduced into the parallel beam of light at right angles to it, some of the light rays are caused to deviate from the parallel bundle and essentially are made to miss the stop. That light is collected through a lens and causes a direct image of the sound either on film or on a television vidicon.

The type of information that schlieren visualization makes available is exemplified by Figure 4 in which three

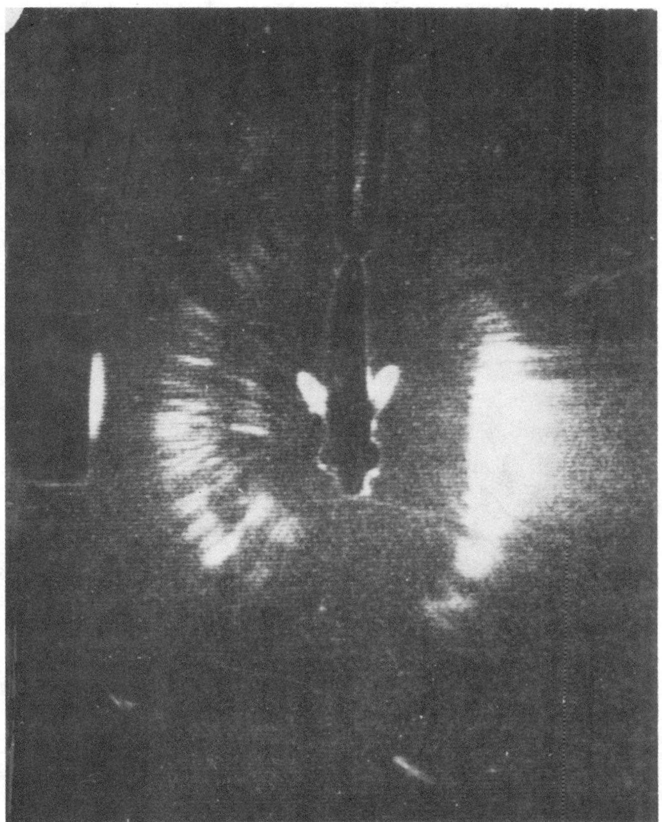

Figure 4. Schlieren photograph of reflection by a fish at 5 MHz. The pulse source is at the left. The outer most reflected wavefront comes from the flesh. The middle wavefront originates at the fish's long skull and the small echo beside the fish comes from the fish's swim bladder. The large white area on the right is pulse energy that bypassed the fish.

different sources of reflection can be identified. They are caused by the flesh, the bony skull and the swim bladder. The reflected wavefronts are very irregular in amplitude, yet their continuity and location can disclose their source. Making pressure field measurements with a hydrophone to obtain the same information would be a formidable, if not impossible chore.

## CREEPING WAVES

The difficulty of slow convergence of special functions for large arguments prompted study of creeping waves in acoustics. It turned out that only a very few terms of a so-called creeping wave series were required for a convergent series to evaluate rigid or soft boundary conditions. However, the real advantage turned out not to be this convergence, but rather the understanding of the individual composite waves that composed reflection and the relationship between them. Waves that result from an interaction of a short pulse with a cylinder are seen in the schlieren picture in Figure 5. The strobed schlieren picture taken at a time when the pulse that was incident on the cylindrical surface has reflected very nearly back to the source at the top of the figure. Transmitted and internally reflected shear (S) and longitudinal (L) waves are indicated in the diagram as a key to the schlieren picture on the left. Extensive work was

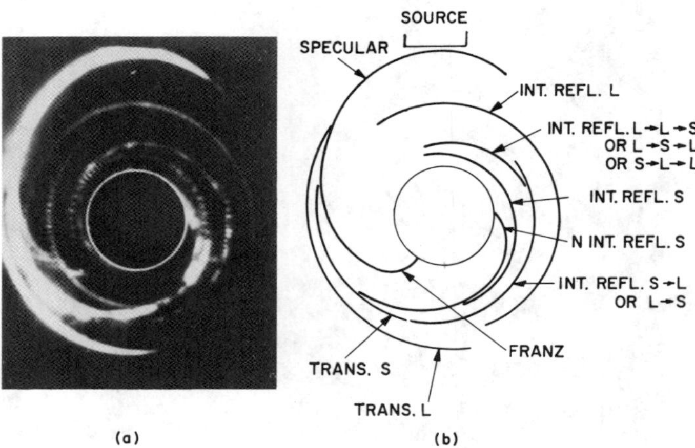


Figure 5. The radiation from an aluminum cylinder at high ka in water resulting from a pulse incident over the top left quadrant. (From W. G. Neubauer and L. R. Dragonette, J. Acoust. Soc. Am. 48, 1135-1149 (1970), with permission of the American Institute of Physics.)

done [5] and despite that, the mechanisms of creeping wave generation in the shadow region are not well understood, nor are they understood for real elastic materials and especially not for absorbing ones.

RESONANCE FORMULATION

Another similar development was the application of so-called resonance methods in acoustic reflection which essentially allowed the combination of geometric and elastic behavior to be described in a single formulation. Since the resonances of a body have to be known beforehand in order to apply the resonance method, again the computation has not yet been significantly simplified. But once again the advantage has been in understanding that was gained. Figure 6 shows an example of that understanding where the backscattered reflection amplitude from a rigid sphere is shown in the upper curve shows the backscattering from a tungsten carbide sphere,

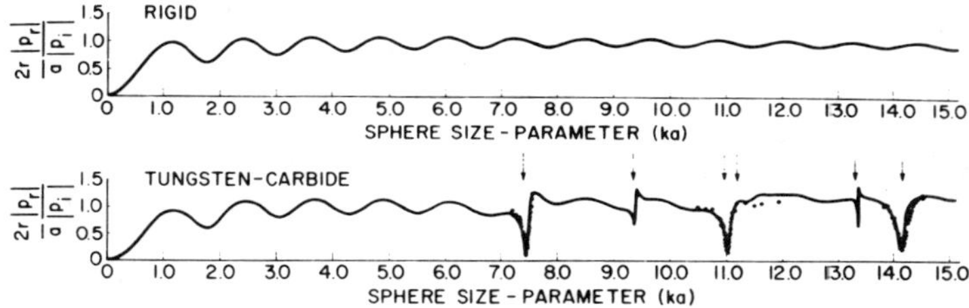


Figure 6. The backscattered reflection from a rigid and a tungsten carbide sphere.

which is very nearly the most rigid material one can find to use in a water medium because of its high specific acoustic impedance. The large deviations indicated by the arrows are positions of ka where specific resonances can be identified for that material. So it appears that the tungsten carbide sphere acts like a rigid sphere except in the region of these resonances. I should mention here that there is a region that is little understood, and that is between the so-called Rayleigh region where there is a predominant volume dependence of the scattering and the resonance region in an elastic body at ka lower than the ka of the gravest resonance.

PLATES AND SHELLS

Another closely allied problem that has a speckled history is the behavior of plate waves. The acoustic reflection from plates, and therefore shells, is not completely in hand although recent strides have occurred for cylindrical shells.

Figure 7 shows a sequence of schlieren pictures of all the plate waves that can be generated on a plate from normal incidence to beyond the critical angle where no transmission occurs.

Figure 8 shows computations by Dragonette [6] which shows the reflection for a thin cylindrical shell as a function of ka. The ratio of inner to outer shell radius is 0.99 so the shell is very thin. At low ka the cylindrical shell acts like a geometric target that is essentially soft, and therefore inverts the reflected pressure wave relative to the incident wave. At the sharp dips, at low ka specific resonant circumferential waves are excited which travel around the circumference of the cylinder and when they are exactly in phase their sum total essentially cancels out the specular reflection. This shell wave has been identified with the first symmetric Lamb mode. As the shell becomes thicker and thicker relative to the incident wavelength as ka is increased the background reflection caused by the geometry changes phase and the negative peaks at low ka change to positive peaks at higher ka because of interference that is now constructive.

In another calculation whose results are shown in Figure 9 the relation between geometric and resonant effect can be seen. The first pulse is the geometric return, out of phase with the incident wave which is shown at the far right. The sequence of smaller pulses are backscattered radiations after each successive transit around the cylinder. It is this sequence that will all be in phase at a resonant point, and for a long pulse length (steady state really) their sum exactly cancels the specular pulse of opposite phase and causes the first zero in Figure 8 at about ka = 3 and subsequent dips.

Often there is little correspondence between computations of boundary value problems giving the possible solutions to the homogeneous equation and observations. That is, plate waves will be computed that have never been observed. And similarly, plate waves are observed that have never been computed. An example of what seems to be a noticeable gap in our understanding is the computation of the relative amplitudes of symmetric and asymmetric Lamb modes in a plate and the behavior of fluid loaded plates.

## PLANE SURFACES

One need not go to a complicated plate in order to find things that have only recently been understood. For example, the reflection of a finite acoustic beam from a homogeneous solid surface in a fluid at the so-called Rayleigh angle had been thought to be displaced, intact, along the surface until

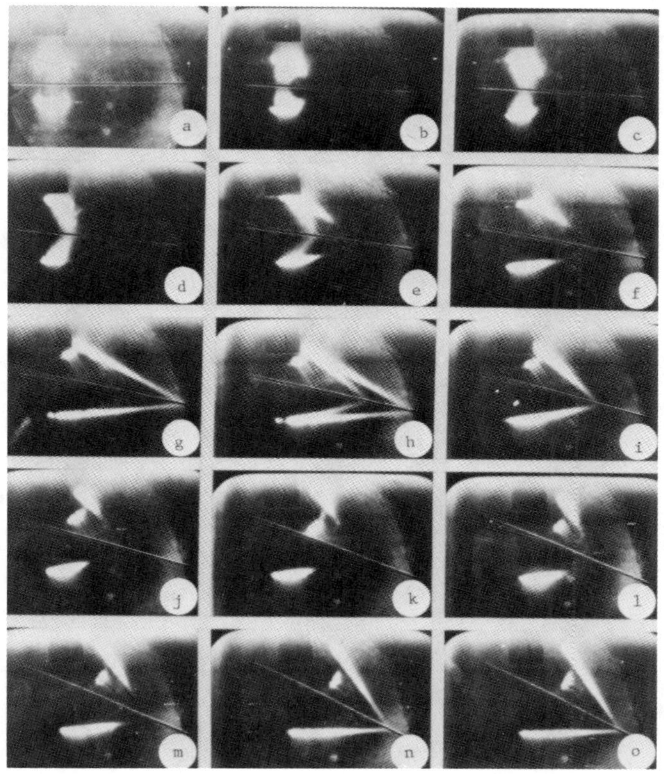


Figure 7. Schlieren photographs taken as a 0.130-cm-thick plate is rotated from normal to the direction of propagation of an incoming acoustic pulse through 26.5°. Pulse length is 3 μsec; frequency is 7.2 MHz. Frame (a) is 0.0°, (b) 2.0°, (c) 4.0°, (d) 6.0°, (e) 8.0°, (f) 10.0°, (g) 12.0°, (h) 12.5°, (i) 13.2°, (j) 16.0°, (k) 20.0°, (l) 22.0°, (m) 24.0°, (n) 26.0°, and (o) 26.5°. (From L. R. Dragonette, J. Acoust. Soc. Am. 51, 920-935 (1972), with permission of the American Institute of Physics.)

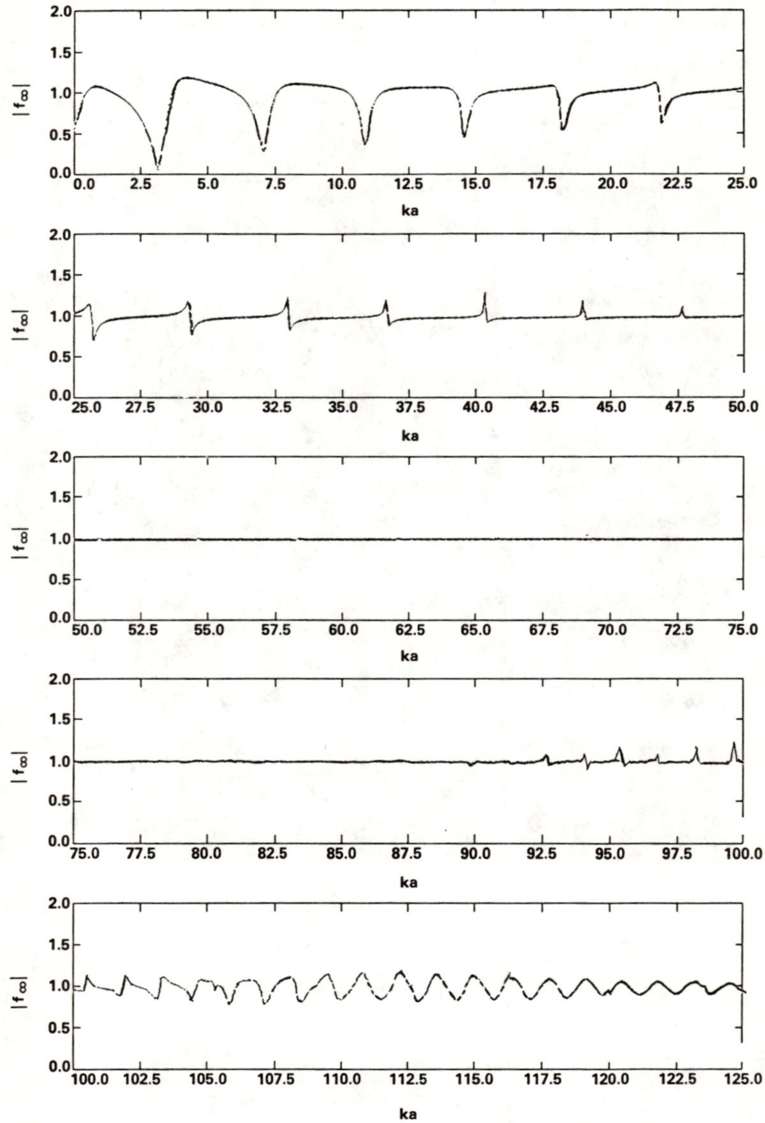


Figure 8. The form function as a function of ka for an aluminum cylindrical shell with a ratio of inner to outer radius of 0.99.

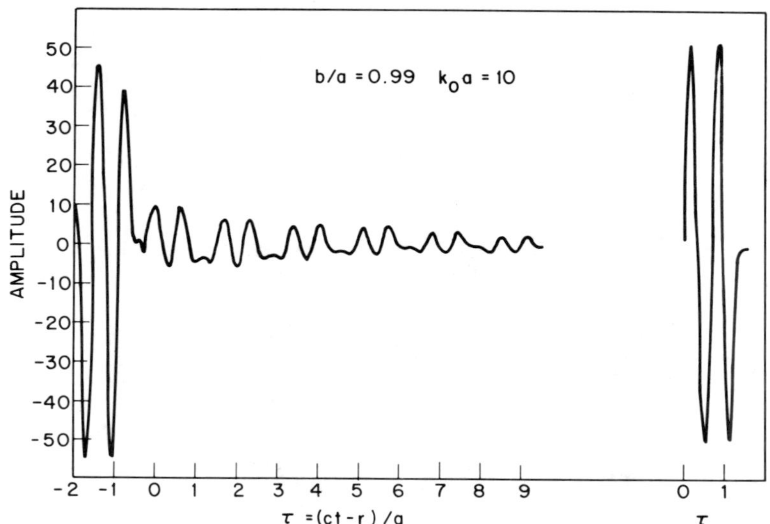


Figure 9. Computation of reflected echoes of a 2 cycle incident pulse shown at the right for a thin cylindrical shell.

fairly recently. This is an area of interest to both body-scattering computations as well as nondestructive evaluation applications.

In Figure 10, the diagram at the top is what was thought to have been physical fact when a wave incident from the left, at the Rayleigh angle on aluminum, is reflected. It was said to have been displaced by a distance delta along the surface. The physical situation that actually occurs [7] is shown on the bottom, where it can be seen that there is a redistribution of energy in the reradiated field when the finite incident beam is incident from the left at the Rayleigh angle. Notice the dark strip occurring in the reradiation. It is not as obvious for the reflection from aluminum as it might be for other materials where a much larger beam displacement is anticipated as computed by earlier theories, such as for aluminum oxide which is shown in Figure 11. Again the picture on the top is the anticipated beam displacement and the observed field is on the bottom.

From these observations and phase measurements in the redistributed reflected field, the description described by Figure 12 was derived. For a finite incident beam, at the Rayleigh angle, a specular beam occurs in the normal position. At the same time, a Rayleigh wave is generated by phase

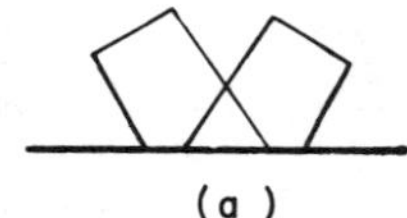

(a)

(b)

Figure 10. Finite beam of 7 MHz ultrasound incident from water on a plane aluminum (1100) interface at the Rayleigh angle. (a) Diagram showing the entire reflected field on the right predicted by beam-displacement theory that should result from the incident beam on the left. (b) Schlieren photograph of an incident beam the same size as that shown in (a) and the resulting reflected and radiated field. (From Reference 7 with permission of the American Institute of Physics.)

matching over the incident area and is reradiated into the fluid as it propagates along the surface. The specular beam and the Rayleigh radiation are out of phase and where they are of equal amplitude they cancel causing the characteristic dark strip between the specular reflection and Rayleigh radiation. After these observations were made, Bertoni and Tamir [8], as well as other people, have actually computed the redistribution of this field. However, the total picture is still incomplete. There are still simple situations such as this that defy understanding and are not embodied in a theoretical description. An example is again at the Rayleigh angle of incidence the finite beam reflects energy back in the direction of the incident wave [9]. That can be seen in the schlieren picture of Figure 13. A more complete outline of the peculiarly shaped backscattered pulse can be seen in

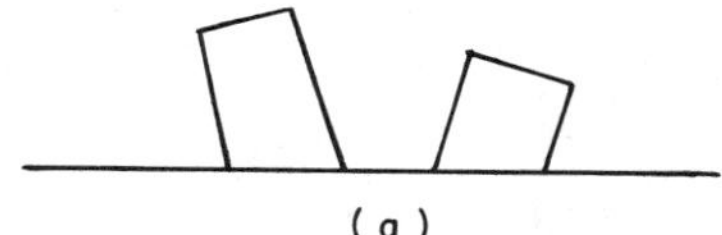

(a)

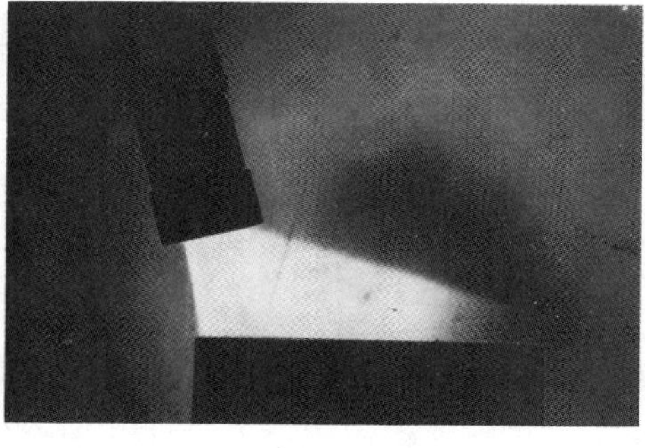

(b)

Figure 11. Finite beam of 7 MHz ultrasound incident from water on a plane aluminum oxide interface at the Rayleigh angle. (a) Diagram showing the entire reflected field on the right predicted by beam-displacement theory that should result from the incident beam on the left. (b) Schlieren photographs of an incident beam the same size as that shown in (a) and the resulting reflected and radiated field. (From Reference 7 with permission of the American Institute of Physics.)

Figure 14, where hydrophone pressure measurements were used to supplement the schlieren visualization. Additional, little understood phenomena seem to occur at both the shear and longitudinal critical angle. In Figure 15 the reradiated field exactly at and near angles slightly above and below the shear critical angle in a high tungsten-content alloy is shown. In Figure 16 a similar observation is shown for the longitudinal critical angle of Pyrex glass.

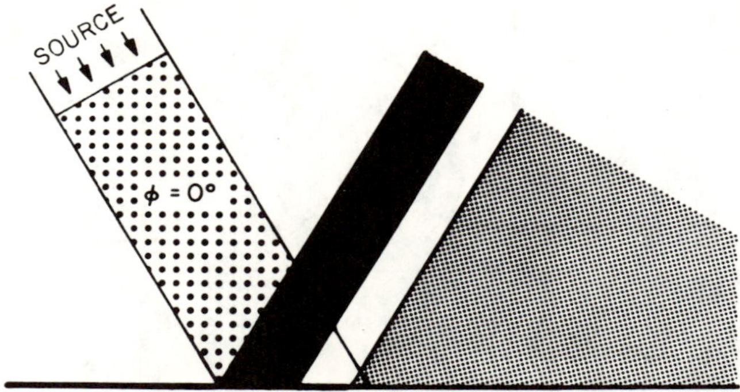


Figure 12. A diagram showing the elements of the reradiated field to the right resulting from the incident beam on the left.

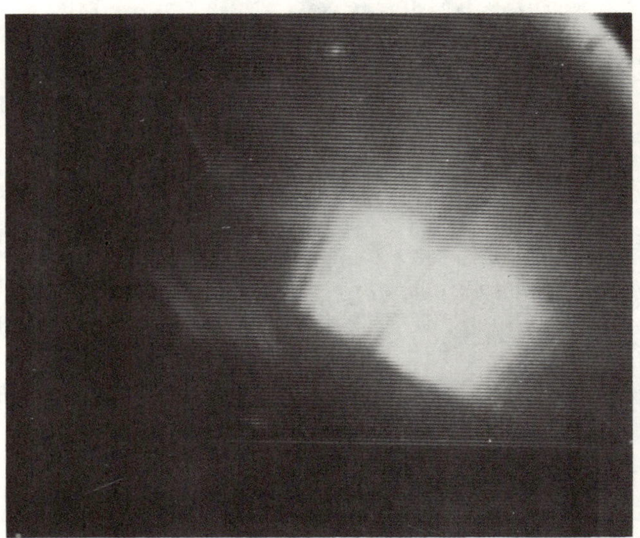

Figure 13. Schlieren photograph of the reradiated field after reflection of a 5 MHz pulse at the Rayleigh angle showing the backscattered pulse.

Figure 14. The photographs of Figure 13 showing the relative phase in the specular and Rayleigh portions of the reradiated field and outlining the back-scattered pulse as measured by an acoustic pressure sensor.

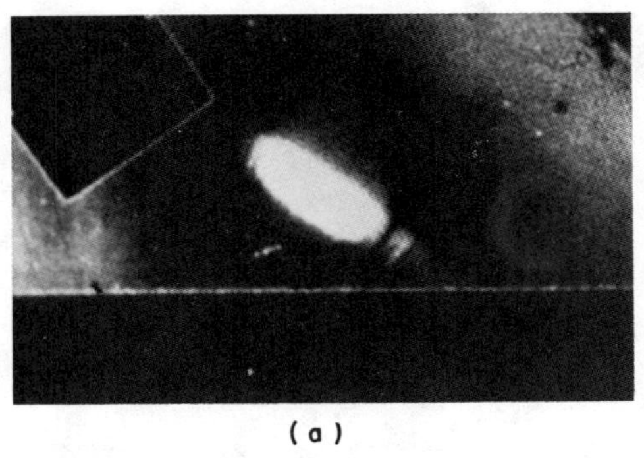
(a)

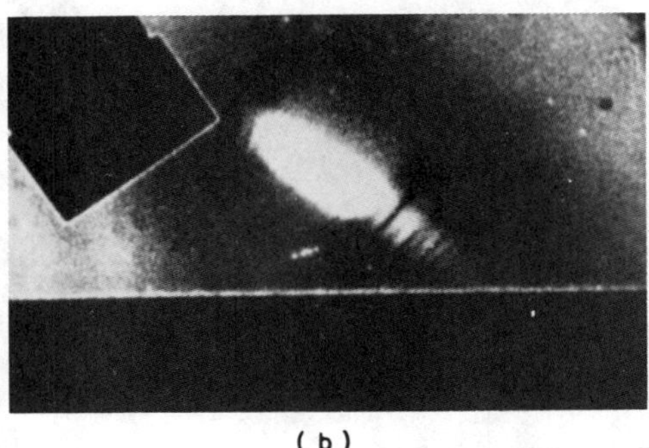
(b)

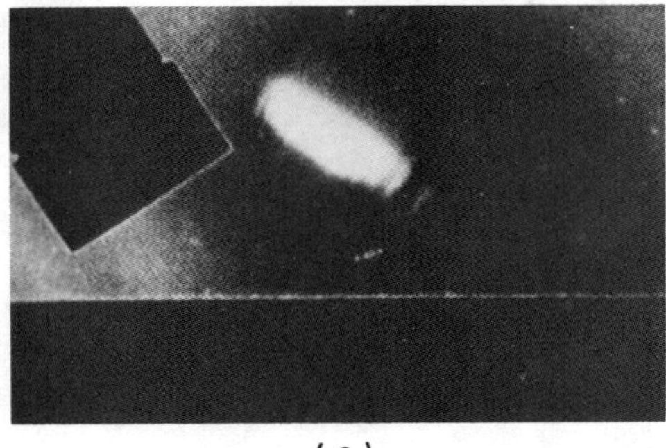
(c)

Figure 15. Schlieren photographs after reflection of a 5 MHz pulse incident from the left on a plane Pyrex-glass-water interface at (a) the incident angle $\theta<15.2^{\circ}$, (b) $\theta = 15.2^{\circ}$ the longitudinal critical angle, and (c) $\theta > 15.2^{\circ}$. (From Reference 7 with permission of the American Institute of Physics.)

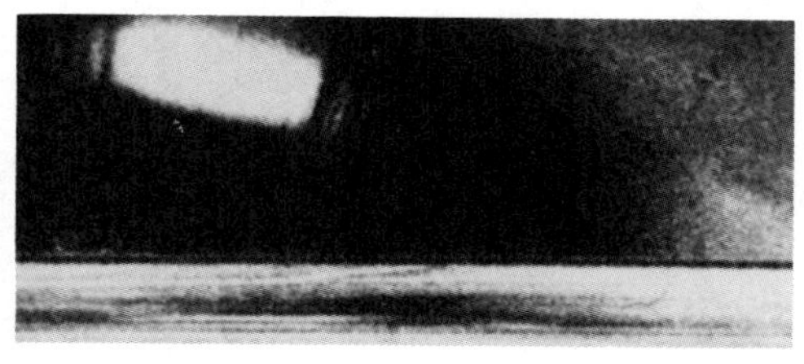

(a)

(b)

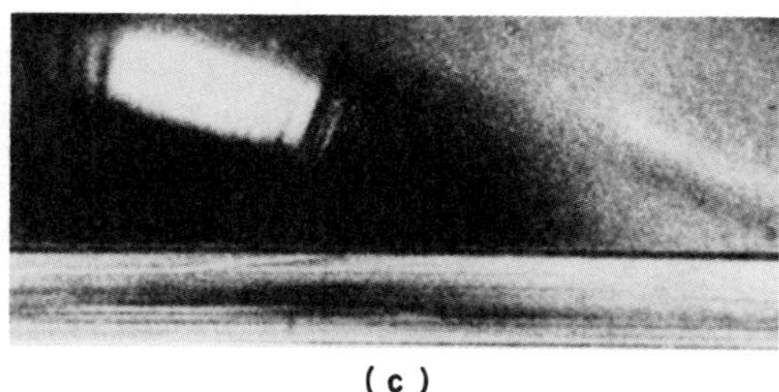

(c)

Figure 16. Schlieren photographs after reflection of a 7 MHz pulse incident from the left on a plane Mallory-1000-water interface at (a) the incident angle $\theta<32.1^{o}$, (b) $\theta = 32.1^{o}$ the shear critical angle, and (c) $\theta > 32.1^{o}$. (From Reference 7 with permission of the American Institute of Physics.)

REFERENCES

1. Fussel, L., Jr. "Basic Problems in Underwater Acoustics Research," Report of the Panel on Underwater Acoustics, (National Research Council, Committee on Undersea Warfare, 1948).

2. Faran, J. J. "Sound Scattering by Solid Cylinders and Spheres," *J. Acoust. Soc. Am.* 23(4):405-418 (1951).

3. Hickling, R. "Analysis of Echoes from a Solid Elastic Sphere in Water," *J. Acoust. Soc. Am.* 34(10):1582-1592 (1962).

4. Flax, L. "High ka Scattering of Elastic Cylinders and Spheres," *J. Acoust. Soc. Am.* 62(6):1502-1503 (1977).

5. Neubauer, W. G., Ugnicius, P. and Uberall, H. "Theory of Creeping Waves in Acoustics and Their Experimental Demonstration," *Z. Naturforsch.* 249(5):691-700 (1969).

6. Dragonette, L. R. Naval Research Laboratory, Report No. 8216 (1978).

7. Neubauer, W. G. "Ultrasonic reflection of a bounded beam at Rayleigh and critical angles for a plane liquid-solid interface," *J. Appl. Phys.* 44:48-55 (1973).

8. Bertoni, H. L. and Tamir, T. "Unified Theory of Rayleigh-angle Phenomena for Acoustic Beams at Liquid-Solid Interfaces," *Appl. Phys.* 2(4):157-172 (1973).

9. Neubauer, W. G. "Acoustic Backscattering in Water at the Rayleigh angle," *J. Acoust. Soc. Am.* 68(S1):S63 (1980).

CHAPTER 2

# A NEW BOUNDARY INTEGRAL EQUATION MODEL FOR THE SCATTERING OF ELASTIC WAVES BY CRACKS

*L. W. Schmerr*

Ames Laboratory, DOE
Department of Engineering Science and Mechanics and the Engineering Research Institute
Iowa State University
Ames, Iowa 50011

## ABSTRACT

This paper describes a new crack model, based on Boundary Integral Equation (BIE) techniques, for solving ultrasonic scattering problems in the low to moderate frequency range where most ultrasonic testing is currently performed. The model treats cracks as zero volume defects where the fundamental unknown is the crack-opening displacement (COD).

This COD-BIE formulation will be solved numerically for the case of antiplane shear waves incident on an isolated or surface-breaking two-dimensional flat crack (slit) and compared with similar results obtained by other methods.

## INTRODUCTION

A crack is often the most dangerous type of flaw to be found in a structural component. Consequently, there is considerable interest in using non-destructive evaluation (NDE) techniques, such as ultrasonics, to detect and characterize cracks during the early stages of their growth. However, in order to make such ultrasonic inspections quantitative, a fundamental understanding of how cracks scatter the ultrasonic waves incident on them is essential.

Recently, we have developed a new crack scattering model, based on Boundary Integral Equation (BIE) techniques, that appears to offer significant advantages for solving ultrasonic scattering problems in the low to intermediate frequency range where most ultrasonic testing is performed. Unlike many previously developed crack models, this new model treats cracks as zero volume defects where the fundamental unknown appearing in the model is the crack-opening displacement (COD). Having a model based directly on the COD avoids the difficult (and often numerically inaccurate) treatment of the crack geometry as the limiting case of a "thin" void. In addition, at low frequencies the COD can be directly related to stress intensity factors which can be used to estimate the structural significance of the crack.

This new COD-BIE formulation will be derived here for the diffraction of horizontal-shear (SH) waves from an arbitrary two-dimensional isolated crack. Numerical results will be obtained for both isolated and surface-breaking planar cracks and compared with some previous crack studies.

## CRACK SCATTERING MODELS

In recent years, there has been considerable interest in elastic wave scattering calculations involving cracks. Mal [1,2], using an integral equation approach, considered the scattering of P and SV-waves from 2-D planar and 3-D penny-shaped cracks and, more recently, for SH-waves incident on a surface-breaking crack [3]. Tan [4], using BIE techniques, developed a model for a planar 2-D crack where the COD's and their derivatives appear as the fundamental unknowns. Using sine and cosine expansions to represent these unknowns, Tan solved the resulting integral equations numerically. Achenbach et al. [5,6], using elastodynamic ray theory, considered the high frequency diffraction of both slits and penny-shaped cracks. More recently, Visscher [7] developed a least-squares boundary residual method for solving the scattering of P and SV-waves by axially symmetric cracks and discussed [8] some of the scattering properties inherent in planar crack models. It should be pointed out that all of the above models (except for SH-wave cases) have been for isolated cracks. To date, the surface-breaking or near-surface crack problem has not been extensively explored, although results are available for a 2-D edge crack [9].

## MODEL FORMULATION

Consider a plane SH-wave obliquely incident on an isolated crack (slit) which has end points located at $\underline{a}$ and $\underline{b}$ (Figure 1). If we model the crack as a line along which we allow a crack opening displacement, $\Delta W$, to occur and assume that under the action of the incident waves the faces of the crack remain stress-free, the waves scattered from the crack can be written in standard integral form as [10]

$$W^S(\underline{P}) = \int_{\underline{a}}^{\underline{b}} [\partial G(\underline{P},\underline{Q})/\partial n'] \quad \Delta W(\underline{Q})\, dS_Q \tag{1}$$

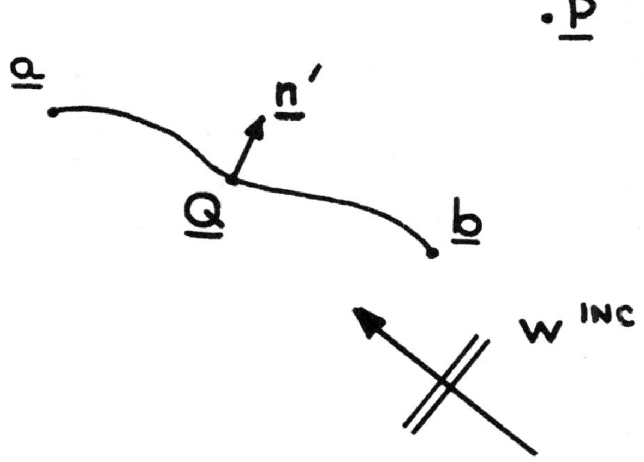


Figure 1. Crack scattering geometry.

where $W^S$ is the displacement of the scattered waves, G is the free space Green's function for Helmholtz's equation, i.e.

$$G(\underline{P},\underline{Q}) = iH_o^{(1)}[k|\underline{P}-\underline{Q}|]/4$$

Here k is the wavenumber and $H_o$ a Hankel function and $\underline{P}$ and $\underline{Q}$ are points located in the media exterior to the crack and along the crack surface, respectively (Figure 1). The vector $\underline{n}'$ is the unit normal to the crack at point $\underline{Q}$.

From Equation 1 we can also compute the stress in the scattered waves along a plane with normal $\underline{n}$, at $\underline{P}$ as

$$\mu \partial W^S(\underline{P})/\partial n = \mu \int_{\underline{a}}^{\underline{b}} \partial^2 G(\underline{P},\underline{Q})/\partial n \partial n' \; \Delta W(Q) \; dS_Q \qquad (2)$$

where μ is the shear modulus.

In order to derive an integral equation for the unknown crack-opening displacement, ΔW, it is tempting to try to take the limit of Equation 2 as point $\underline{P}$ approaches the crack boundary and use the stress-free conditions of the crack face, i.e.,

$$\mu \partial W^S/\partial n = -\mu \partial W^{INC}/\partial n$$

where $W^{INC}$ is the known displacement of the incident wave, to write

$$-\partial W^{INC}(\underline{P})/\partial n = \int_{\underline{a}}^{\underline{b}} \partial^2 G(\underline{P},\underline{Q})/\partial n \partial n' \, \Delta W(Q) \, dS_Q \qquad (3)$$

where both $\underline{P}$ and $\underline{Q}$ in Equation 3 are on the boundary. Unfortunately, this method fails since the $\partial^2 G/\partial n \partial n'$ integrand is strongly singular for $\underline{P}$ near $\underline{Q}$ and the integral is divergent [11]. However, a convergent formulation can be developed if we note that for $\underline{P}$ not on the boundary of the crack, it can be shown that

$$\int_{\underline{a}}^{\underline{b}} \partial^2 G(\underline{P},\underline{Q})/\partial n \partial n' \, dS_Q$$

$$= \hat{\underline{Z}} \cdot [\underline{n} \times \underline{\nabla} \, G(\underline{P},\underline{Q})] \Big|_{\underline{a}}^{\underline{b}} + k^2 \int_{\underline{a}}^{\underline{b}} \underline{n} \cdot \underline{n}' \; G(\underline{P},\underline{Q}) \, dS_Q \qquad (4)$$

where $\hat{\underline{Z}}$ is a unit vector in the Z-direction.

Equation 4 follows directly from the expansion of the left side of Equation 4 in terms of the x and y-derivatives of G and using the fact that G satisfies Helmholtz's equation. With this result, we can multiply Equation 4 by $\Delta W(\underline{P})$ and subtract it from Equation 2. Then taking the limit of the resulting equation, as $\underline{P}$ goes to the boundary, gives

$$-\partial W^{INC}(\underline{P})/\partial n = \int_{\underline{a}}^{\underline{b}} \partial^2 G/\partial n \partial n' \, [\Delta W(\underline{Q}) - \Delta W(\underline{P})] dS_Q$$

$$+ \Delta W(P) \; \{\hat{\underline{Z}} \cdot [\underline{n} \times \underline{\nabla} \, G] \Big|_{\underline{a}}^{\underline{b}} + k^2 \int_{\underline{a}}^{\underline{b}} n \cdot n' \, G \, dS_Q\} \qquad (5)$$

where now all the terms are convergent. Equation 5 can be viewed as representing the finite part of the divergent integral in Equation 3. A relation similar to Equation 5 for three-dimensional acoustic scattering problems has been recently given by Terai [12]. This type of divergent integral has also been treated by analogous methods for acoustic scattering problems involving smooth scattering geometries [13],

[14]. However, for 2-D crack-like geometries, this formulation is apparently new.

NUMERICAL RESULTS

Although Equation 5 was derived for the case of an isolated flaw, it can also be used to solve for the scattering of waves from a surface-breaking crack. To see this, consider a plane crack, of width 2a, lying along the x-axis (Figure 2). By subjecting this crack to symmetrical right and leftward propagating waves $W_R$ and $W_L$ (Figure 2), the scattered displacement will, by symmetry, be an even function of x. Thus, the total stress normal to the boundary x=0 (which is proportional to $\partial w/\partial x$) will vanish automatically. The isolated crack problem of Figure 2, therefore, can be used to solve the surface-breaking crack problem, shown in Figure 3, where $W_R$ can be interpreted physically as the plane wave reflected directly from the boundary y=0.

For the geometry of Figures 2 and 3, the terms in Equation 5 can be written in particularly simple form if we break the crack face up into n elements and assume that the crack-opening displacement, $\Delta W$, is a constant over each element. Then Equation 5 becomes

$$-\partial W^{TOT}(x,0)/\partial y = \sum_{i=1}^{n} \Delta W_i \left\{ \partial G(k|x-x'|)/\partial x' \Big|_{x'=x_i}^{x'=x_{i+1}} + k^2 \int_{x_i}^{x_{i+1}} G(k|x-x'|)\, dx' \right\} \qquad (6)$$

where $W^{TOT} = W_L + W_R$ is the sum of the displacements of incident and reflected plane waves and $\Delta W_i$ is the crack opening displacement evaluated at the centroid, $x_i$ of the ith element $(x_i, x_{i+1})$. For a plane wave incident at an angle $\theta$ with respect to the x-axis (Figure 3), we have for $W^{TOT}/\partial y$

$$\partial W^{TOT}(x,o)/\partial y = 2\, t_o \sin\theta \cos[kx\cos\theta]/\mu$$

where $t_o$ is the stress amplitude of the incident wave. Placing this expression into Equation 6 and satisfying that equation at the centroid of each element then yields a set of simultaneous equations in the n unknowns $\Delta W_i$ which can be written formally as

$$\sum_{i=1}^{n} A_{mi}\Delta W_i = f_m \quad (m=1.2...n) \qquad (7)$$

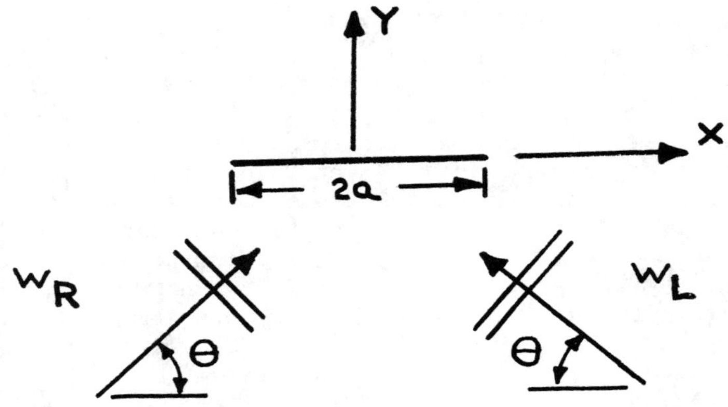


Figure 2. Plane waves incident on a straight crack.

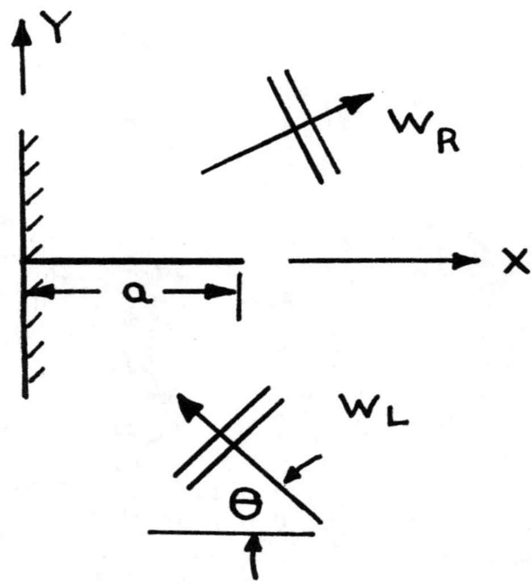


Figure 3. Surface-breaking crack geometry.

where

$$A_{mi} = kH_1^{(1)} (k |\hat{x}_m - x_{i+1}|) \operatorname{sgn} (\hat{x}_m - x_{i+1})$$

$$-k H_1^{(1)} (k |\hat{x}_m - x_i|) \operatorname{sgn} (\hat{x}_m - x_i)$$

$$+k^2 \int_{x_i}^{x_{i+1}} H_o^{(1)} (k|\hat{x}_m - x|) dx$$

and

$$f_m = (8i\, t_o \sin\theta/\mu) \cos [k\, x \cos \theta]$$

$$\operatorname{sgn}(\xi) = \begin{array}{l} +1 \quad \xi>0 \\ -1 \quad \xi<0 \end{array}$$

For m $\neq$ i the integral in the expression for $A_{mi}$ can be performed numerically by standard integration techniques such as Simpson's method, while for m=i, the integrals can be performed exactly, leading to Struve functions that can be evaluated numerically from their series expansions [15].

The results of solving Equation 7 for a plane wave incident on a surface-breaking crack at $10^o$, $45^o$, and $90^o$ are shown in Figures 4-6. For all these cases the crack length 2a contained a total of 50 elements of equal length. The COD distributions along the crack face were plotted in Figures 4-6 for various non-dimensional frequencies, ka. In those figures, the actual COD's were normalized by the maximum static (ka=0) COD, $W_{ST}$, where in our case $W_{ST}$ is given by

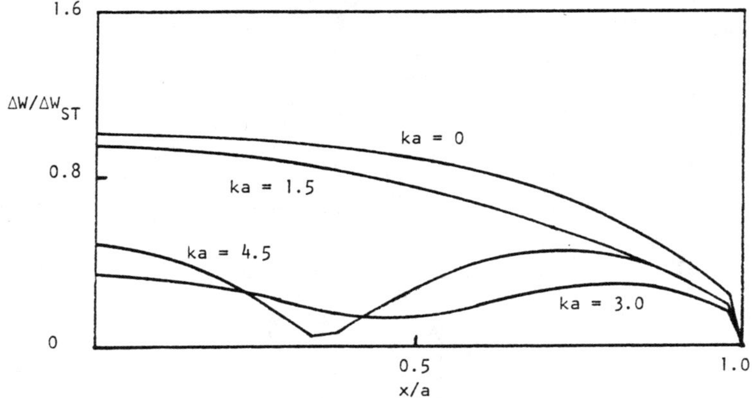


Figure 4. Crack-opening displacement distribution for a plane wave incident at $\theta = 10^o$.

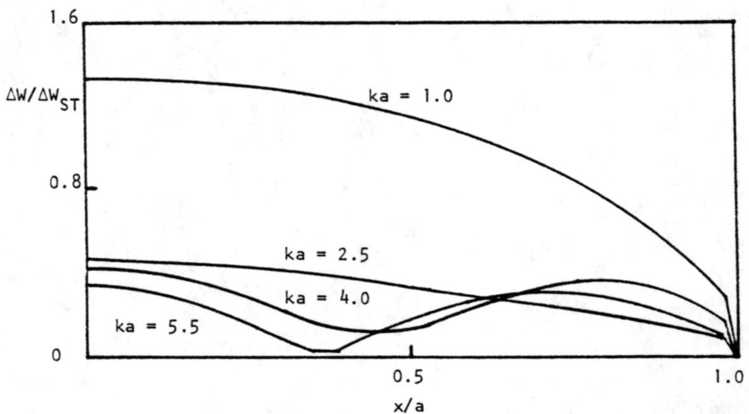


Figure 5. Crack-opening displacement distribution for a plane wave incident at $\theta = 45^o$.

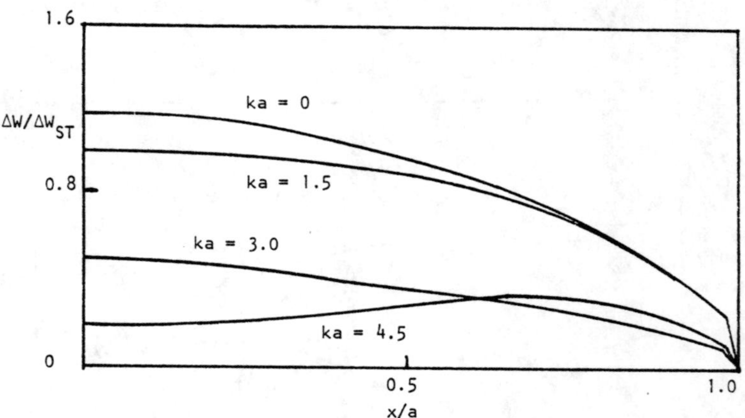


Figure 6. Crack-opening displacement distribution for a plane wave incident at $\theta = 90^o$.

$$W_{ST} = 4t_o \sin\theta/\mu$$

From these results, we can see that for the lowest ka values, the COD distribution is similar to a static COD curve, while at higher frequencies an oscillatory pattern develops across the crack face. These results can be shown to agree closely with those of Mal [1 ,3] , except near the crack tip x=a where the mesh size is obviously inadequate to reproduce the local behavior of $\Delta W$. In fact, near the crack tip x=a , $\Delta w \alpha (a-x)^{1/2}$ so that the slope of $\Delta W$ becomes infinite at the tip. The discrepancy between the numerical and exact result can be easily minimized, however, by using a finer element spacing near the crack tip. It should be pointed out that in the $\theta=90^o$ case, the incident and reflected waves coalesce to form a single wave at normal incidence to the crack face. Thus, the results of this case can be directly compared with those of Mal [1] for an isolated crack.

Having the COD distribution on the crack face in effect solves the scattering problem completely, since as Equations 1 and 2 show, the scattered displacement or stress can then be computed anywhere in the surrounding medium. In particular, the far-field scattering amplitude, which is of interest in many ultrasonic NDE applications, can be calculated directly [3].

## SUMMARY

We have shown how a new COD-BIE crack scattering formulation can be used effectively to solve for the scattering of SH-waves from two-dimensional isolated or surface-breaking cracks. This same type of model can be generalized for P and SV-wave scattering problems for isolated cracks. A description of the method in these cases is forthcoming shortly. The SH-wave surface-breaking crack model is currently being used as a means of testing various approximate methods for handling the much more difficult problem of P and SV-wave scattering by near-surface and surface-breaking cracks. Some methods being explored for these problems include modified Green's function approaches and the use of special elements to model radiation conditions.

## ACKNOWLEDGEMENT

This work was supported by the U.S. Department of Energy, contract No. W-7405-Eng-82, Division of Basic Sciences, Budget Code AK-01-02-01-05.

## REFERENCES


1. Mal, A. K. "Interaction of Elastic Waves with a Griffith Crack," *Int. J. Eng. Sci.* 8(9):763-776 (1970).

2. Mal, A. K. "Interaction of Elastic Waves with a Penny-Shaped Crack," *Int. J. Eng. Sci.* 8(5):381-388 (1970).

3. Stone, S. F., Ghosh, M. L. and Mal, A. K. "Diffraction of antiplane shear waves by an edge crack," *J. Appl. Mech. Trans. ASME* 47(2):359-362 (1980).

4. Tan, T. H. "Scattering of plane elastic waves by a plane crack of finite width," *Appl. Sci. Res.* 33(1):75-88 (1977).

5. Achenbach, J. D., Gautesen, A. K. and McMaken, H. In: *Acoustic, Electromagnetic and Elastic Wave Scattering,* V. K. Varadan and V. V. Varadan, Eds. (New York: Pergamon Press, 1980), p. 355.

6. Achenbach, J. D., Gautesen, A. K. and McMaken, H. In: *Modern Problems in Elastic Wave Propagation*, J. Miklowitz and J. D. Achenbach, Eds. (New York: John Wiley, 1978), p. 219.

7. Visscher, W. M. "Scattering of Elastic Waves from Planar Cracks in Isotropic Media," *J. Acoust. Soc. Am.* 69(1): 50-53 (1981).

8. Visscher, W. M. "Calculation of the Scattering of Elastic Waves from a Penny-Shaped Crack by the Method of Optimal Truncation," *Wave Motion* 3:49 (1981).

9. Achenbach, J. D., Keer, L. M. and Mendelsohn, D. A. "Elastodynamic Analysis of an Edge Crack," *J. Appl. Mech. Trans. ASME* 47(3):551-556 (1980).

10. Pao, Y. H. and Mow, C. C. *Diffraction of Elastic Waves and Dynamic Stress Concentrations* (New York: Crane-Russak, 1972).

11. Filippi, P. J. T. "Layer Potentials and Acoustic Diffraction," *J. Snd. Vib.* 54(4):473-500 (1977).

12. Terai, T. "On Calculation of Sound Fields Around Three-Dimensional Objects by Integral Equation Methods," *J. Snd. Vib.* 69(1):71-100 (1980).

13. Meyer, W. L., Bell, W. A. and Zinn, B. T. "Boundary Integral Solutions of Three Dimensional Acoustic Radiation Problems," *J. Snd. Vib.* 59(2):245-262 (1978).

14. Sayhi, M. N., Ousset, Y. and Verchery, G. "Solution of Radiation Problems by Collocation of Integral Formulations in Terms of Single and Double Layer Potentials," *J. Snd. Vib.* 74(2):187-204 (1981).

15. Abramovitch, M. and Segun, I. A. *Handbook of Mathematical Functions* (New York: Dover Publications, 1965).

CHAPTER 3

# ACOUSTIC DIFFRACTION FOR HIGHWAY NCISE BARRIERS

*Sabih I. Hayek**

The Department of Engineering Science and Mechanics and
The Applied Research Laboratory
The Pennsylvania State University
University Park, Pennsylvania 16802

The problem of reducing roadside traffic noise in urban areas has been solved by erecting noise barriers. These barriers are constructed from natural or man-made materials which reduce the directly transmitted sound appreciably. However, some of the acoustic energy is diffracted over the barrier or scattered by other objects, so that part of the acoustic energy is deflected to homes behind the barrier. The effectiveness of a barrier, measured by its insertion loss, depends on the height of the barrier, its shape and whether it has sound absorbing coverage. Models for the acoustic diffraction for different absorbent or rigid barriers are presented and typical predictions are given for a barrier. Laboratory 1:5 scale model experiments were made and typical results for such tests are shown.

## INTRODUCTION

Highway noise in urban areas is one area of noise pollution that can be rectified. The primary mechanism for traffic noise reduction in existing highways is through the use of barriers. Noise barriers are physical barriers that can reduce the noise transmitted directly from the vehicles in the traffic. They can be constructed from natural materials, such as earth and wooden barriers or from man-made materials such as fiberglass, aluminum, concrete, etc.

To define the noise problem first, one needs to talk about traffic noise. Vehicles emit noise due to different causes. For example, an automobile has engine noise and tire generated noise. While some of these noise sources are pure

*Currently on leave at the Naval Ocean Systems Center, Code 635, San Diego, CA 92152

tone, the overall radiated noise can be treated as a random noise with a known spectrum. Because the dominant noise from an automobile comes from the vicinity of the tires, the overall noise source height is located at the ground, with a typical normalized noise spectrum shown in Figure 1. Similar arguments are advanced for the noise emitted from light trucks. On the other hand, heavy trucks have a variety of noise sources. Among the most important are exhaust noise (at a height of 3.6m), engine noise (at a height of 1.2m), tire noise (at zero height), radiated noise from side panels and flow noise from the truck itself. An accepted weighted center of these sources is at 2.4m above ground and has a typical normalized spectrum shown in Figure 1. However, the normalized spectra do not show that a typical heavy truck can emit 25dB more noise than an automobile.

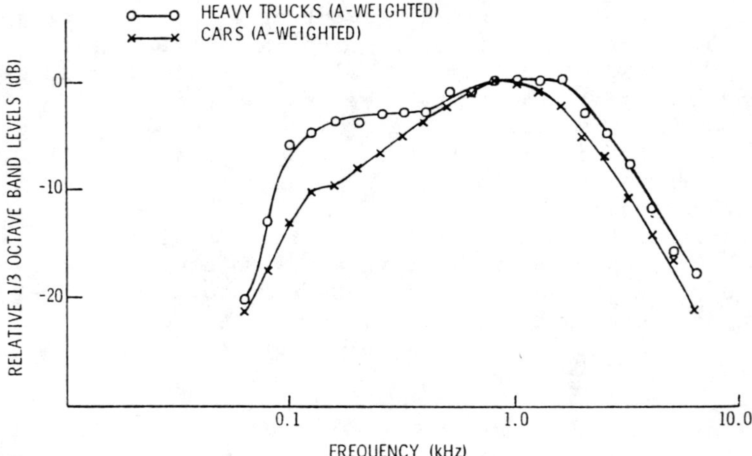


Figure 1. Normalized noise spectra for cars and heavy trucks.

Traffic noise is a composite of a mixture of automobiles, light and heavy trucks. The total noise level received at a point is the sum of the noise in each frequency band of all the vehicles traveling in all the lanes of a highway, and then added incoherently in space and over all frequency bands. This overall traffic noise level is the basis for evaluating the need for, and effectiveness of, a noise barrier. A noise barrier erected between the traffic and the receiver acts to reduce the directly transmitted noise from each vehicle. If the transmission loss of the barrier is high enough (say larger than 20dB) then the only noise that can be received is through the mechanism of acoustic diffraction over the barrier's top or side edge(s). Ideally, one has to recalculate the diffracted energy from each vehicle in each lane in each frequency band and then add the contributions incoherently over space and frequency to arrive at an overall level at the receiver site. The ratio in dB of the received level due to the traffic to that after the erection of a barrier is called the Insertion Loss (EL). The ratio of the level received as if the traffic lines are free in space to that received after the erection of a barrier is called the Excess Attenuation (EA). The difference between them is that the former is more realistic since it takes into account the acoustic rays reflected off the pavement in calculating the traffic noise level before and after the erection of a barrier. The latter assumes that the traffic noise line is located in free space, so that no reflections are accounted for noise level calculations before and after the erection of a barrier. The level of traffic noise before and after an erection of a noise barrier requires that one knows the precise mix of the three types of vehicles and the order they appear at any instant in time. This requirement is certainly very severe, requiring a lot of computational effort. A more accepted method is to assume that the traffic noise is due to a uniform distribution of vehicles of one type, all having identical spectra and whose noise energies added incoherently. Thus, one can compute barrier attenuation for each type of vehicle and then compute an overall number that corresponds to a typical average day mix of vehicles.

## FREE FIELD COMPUTATION

The noise power spectrum of a vehicle can be divided into 17 to 21 1/3-octave bands and the total radiated power of a vehicle is divided by the vehicle spacing to give a uniform power distribution per unit length of a finite or an infinitely long traffic line. The total noise received at any receiver is then integrated over the length of line of the traffic by summing incoherently the noise power generated by line segments of equal angular intervals in each frequency band and then compute the total field by summing incoherently over the number of frequency bands considered. For example, the summation is carried out for equal intervals $d\gamma$ at each central location $\gamma$, with the summation carried over the interval $(\gamma_1, \gamma_2)$ as shown in Figure 2.

## NOISE BARRIERS

Noise barriers come in a variety of shapes. A noise

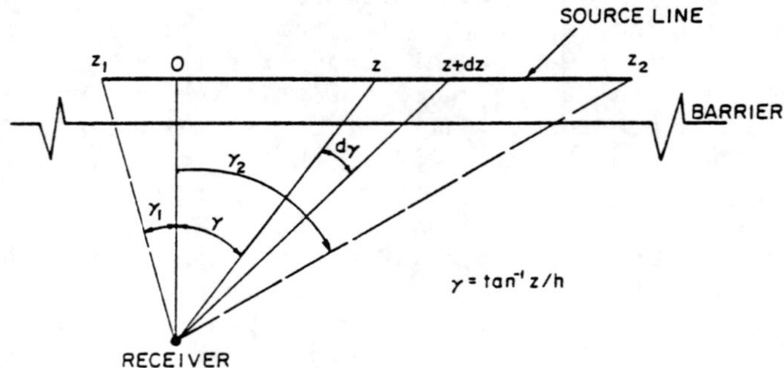


Figure 2. Integration scheme for an incoherent line.

control highway engineer has to take into consideration the cost of materials and construction, the esthetic appearance of the barrier, its acoustic absorbency, its resistance to environmental conditions, durability, ease of repair and maintenance, as well as its acoustic performance. These factors influence the choice of the barrier's shape, height, length and its location in relation to the pavement. Typical shapes for highway noise barriers are shown in Figure 3. Some of the massive barriers, such as trapezoids and wedges, can be shaped inexpensively from earth excavated during construction and shaped for ease of maintenance. These have good transmission loss characteristics and are usually planted with vegetation. Other barriers, such as the thin wall, T and L shaped, are man-made and thus, material and construction cost per unit length is high.

ACOUSTIC DIFFRACTION MODELS

To predict the effectiveness of a barrier design, an analytic model for the diffracted acoustic energy is needed. The total field p at a receiver point is calculated

$$p = p_i + p_r + p_d$$

where $p_i$ is the pressure at the receiver due to a point source located along a line, $p_r$ is the reflected pressure and $p_d$ is the diffracted pressure. The incident field is calculated for a point source whose strength is equivalent to the power generated from a segment of an incoherent line source. The reflected pressure is calculated by use of standard

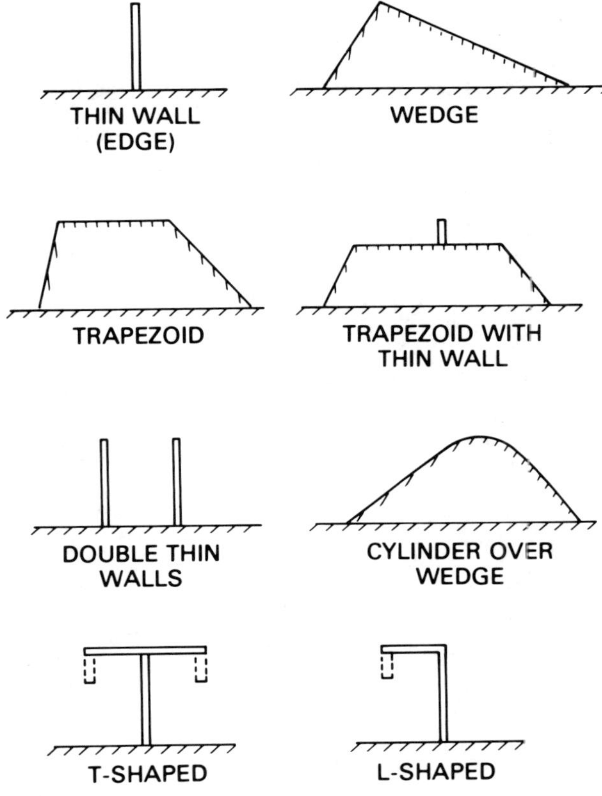


Figure 3. Typical noise barrier configurations.

reflection coefficients. The diffracted pressure is calculated from an exact analytic model or in many practical cases by use of the Geometrical Theory of Diffraction (GTD) [1] together with some canonical analytic exact or approximate solutions.

Typically, for a singly diffracted acoustic ray, the ray travels from the source to a point on the barrier and then to a receiver, see Figure 4, for a thin-wall barrier. The total pressure is given by:

$$
\begin{aligned}
P &= P_d && \text{Region (I)}, \; -\pi \le \phi \le -\pi+\phi_o \\
&= P_d + P_i && \text{Region (II)}, \; -\pi+\phi_o \le \phi \le \pi-\phi_o \\
&= P_d + P_i + P_r && \text{Region (III)}, \; \pi-\phi_o \le \phi \le \pi \; ,
\end{aligned}
$$

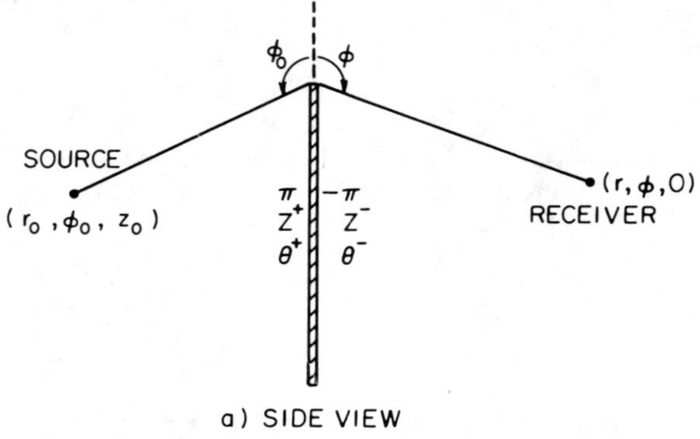


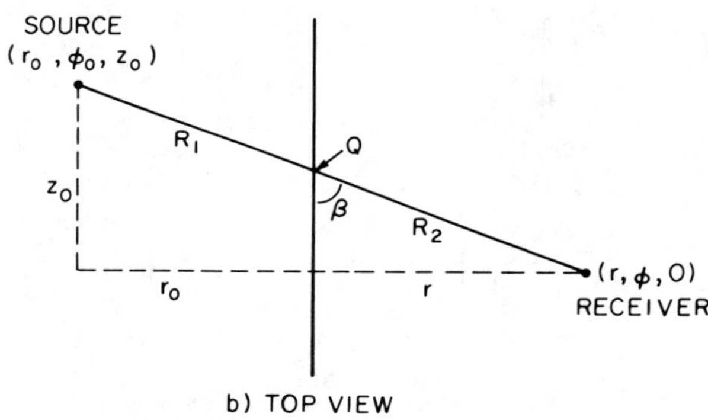


Figure 4. Geometry for diffraction from a thin-wall.

where the regions are depicted in Figure 5. The GTD [1] solution for the diffracted ray is:

$$P_d = P_i\ (Q)\ A(R_1,\ R_2)\ D(\phi,\ \phi_o,\ \beta)\ e^{-ikR_2} \tag{2}$$

where $P_i$ (Q) is the incident pressure at point Q on the barrier, D is the diffraction coefficient that corresponds to the particular shape of the barrier, $A(R_1, R_2)$ is the amplitude factor that depends on the radii of curvatures of the diffracted wave front and the last term is a phase term for the distance from the point Q on the barrier to the receiver. Typical expressions for $p_d$ for a thin-wall barrier are enumerated below [2]:

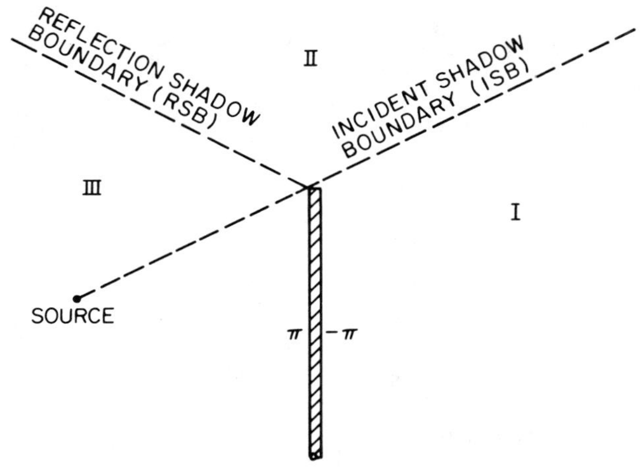


Figure 5. The three regions for the thin-wall barrier.

$$D(\phi,\phi_o,\beta) = \frac{\text{sgn (A)} \sqrt{kL}\ e^{+iA^2}}{\sqrt{2k}\ \ 2\ \sin\beta} [1-\sqrt{2}\ \ e^{i\pi/4}\ F^*(|A|)]$$

$$+ \frac{\text{sgn (B)} \sqrt{kL}\ e^{+iB^2}}{\sqrt{2k}\ \ 2\ \sin\beta} [1-\sqrt{2}\ \ e^{i\pi/4}\ F^*(|B|)], \qquad (3)$$

where sgn(x) is the sign of the argument,

$$A = \sqrt{kL}\ \cos\frac{\phi + \phi_o}{2},\ B = \sqrt{kL}\ \cos\frac{\phi - \phi_o}{2},$$

$$L = \frac{2\ R_1R_2}{R}\sin^2\beta,\ F^*(x) = C(x) - iS(x),$$

$$C(x) = \sqrt{\frac{2}{\pi}}\int_o^x \cos t^2\ dt,\ S(x) = \sqrt{\frac{2}{\pi}}\int_o^x \sin t^2\ dt,$$

$$A(R_1,R_2) = \sqrt{R_1/RR_2}\ ,\ P_i = \frac{e^{-ikR_1}}{R_1},\ \text{and } R = R_1 + R_2\ .$$

If the barrier is shaped such that some rays may be diffracted twice, such as a double-wall, trapezoid, etc., then the diffracted pressure may be found by use of the GTD, where exact or approximate analytic solutions may not be available. Typical of these doubly diffracted pressure calculations is the one for a double-thin-wall (See Figure 6):

$$P_d = P_i\ (Q)\ D(\phi,\phi_o,\beta)\ D(\phi',\phi_o',\beta)\ A(R_1,R_2,R_3)\, e^{-ik(R_2+R_3)} \tag{4}$$

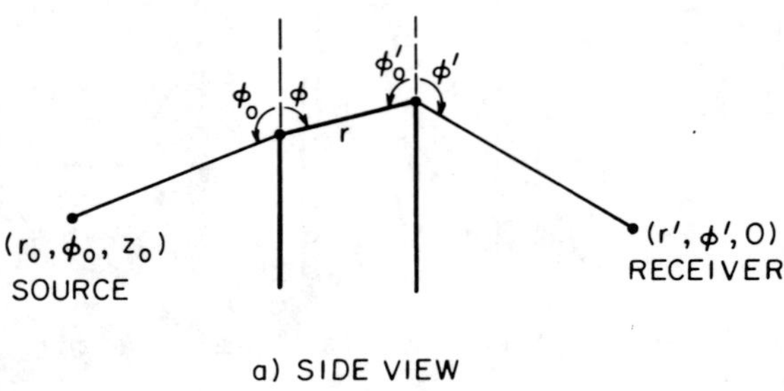


a) SIDE VIEW

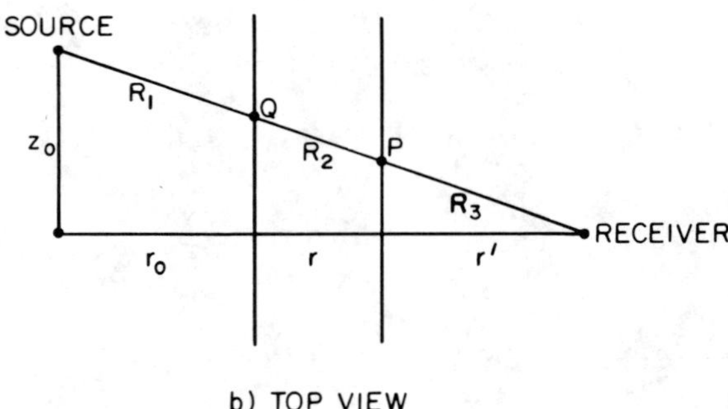


b) TOP VIEW

Figure 6. Geometry for diffraction from a double-wall-wall.

where the amplitude factor is defined by:

$$A(R_1,R_2,R_3) = \sqrt{R_1/R_2\ R_3\ R} \quad , \tag{5}$$

and the diffraction coefficients are calculated for the entry and exit angles and radii.

It is thus essential that one obtains analytic solutions for the specific shape of the barrier. Often, the solution for the diffracted pressure for a plane wave is the only one available. A plane-wave solution can be converted approximately to a solution for a point source by replacing the length distance to the receiver r by the parameter L of Equation 3.

EXCESS ATTENUATION

The diffracted pressure from one segment of an incoherent line source, treated as point source, is calculated for one receiver location at a frequency centered at a 1/3-octave band. This calculation is repeated for all the 1/3-octave bands and summed incoherently (on a power basis). These calculations are again repeated from another point source on the source line and resulting incoherent summation over all the line segments gives the total acoustic power at one receiver point due to an incoherent line source. The ratio of the free field power to the received power with the barrier in place in dB gives the Excess Attenuation for the barrier at the specified receiver point. To get a meaningful and informative picture of the effectiveness of the barriers, it is essential that one performs these calculations over a field of receiver points and contours of equal excess attenuation in dB are plotted. These procedures were used in a theoretical noise barrier's assessment at the Pennsylvania State University [3,4]. Computations were performed for a traffic line source length of 12,000 ft. (3.6km) located at some offset distance from the barrier. The receiver field consisted of 88 receiver points, see Figure 7. Typical contours of excess attenuation for a thin wall of height 15 ft. (4.50m) is shown in Figure 8 for a line of heavy trucks at offset distance of 60 ft. (18m) from the barrier.

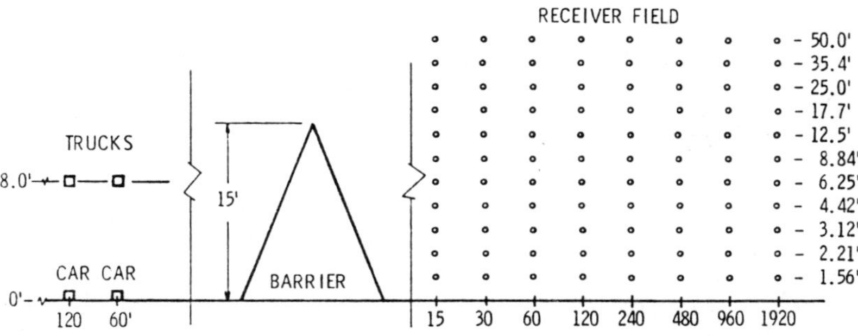


Figure 7. Traffic line source and receiver positions.

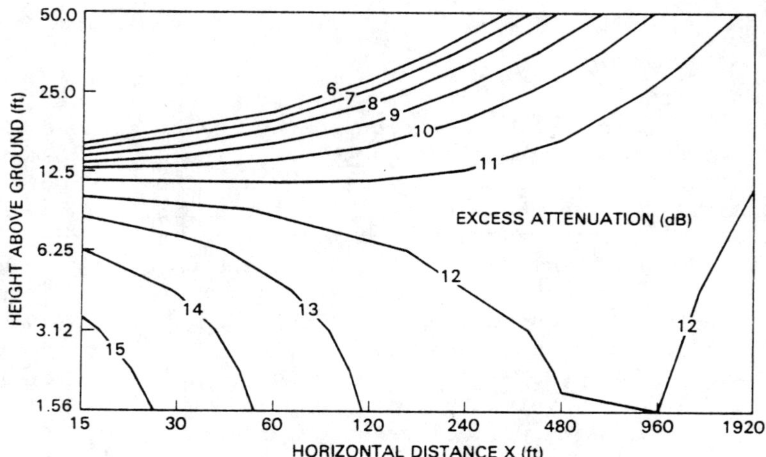


Figure 8. Contours of EA for a heavy truck line source at offset distance of 60 ft.

INFLUENCE OF GROUND

In the foregoing analysis, the source was assumed to be in a free field. The assessment of the efficiency of noise barriers to reduce the traffic noise was thus judged solely on the excess attenuation figures. However, when a source or receiver is located above the ground, the reflected rays and any other rays may become significant. In the calculation of the insertion loss, one must first calculate the noise level existing prior to the erection of a barrier. An approximation can be made, where the total field is composed of the incident and reflected rays (see Figure 9):

$$p = \frac{\exp(ikR_1)}{R_1} + R_p \frac{\exp(ikR_2)}{R_2} , \tag{6}$$

where

$$R_p = \frac{\sin\psi - \beta}{\sin\psi + \beta} , \tag{7}$$

and

$$\beta = \frac{\rho c}{Z} , \tag{8}$$

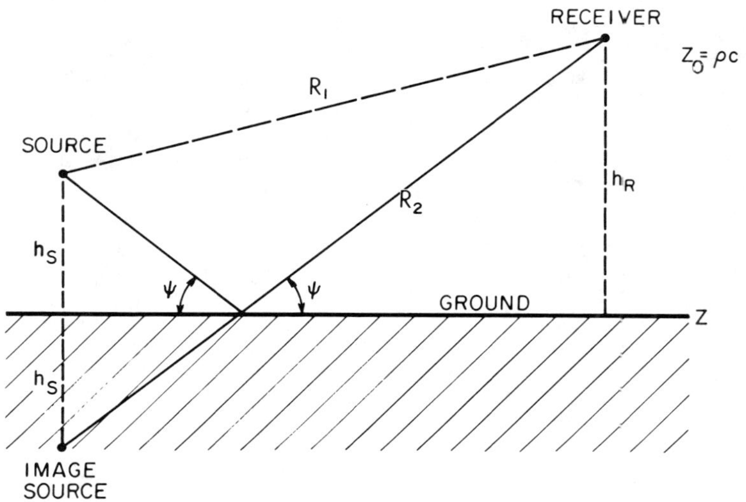


Figure 9. Geometry for diffraction off the ground.

are the plane wave reflection coefficient and normalized admittance of the ground, respectively; ρc, is the air characteristic impedance and Z the surface acoustic impedance. The total field given by Equation 6 would give accurate results when either the source or the receiver is located few wavelengths above the ground. However, when both source and receiver are very close to the ground, then $\psi \to 0$, $R_1 = R_2$, the reflection coefficient $R_p = -1$ so that the total field given by Equation 6 vanishes. This fact led to further investigations into wave propagation near the ground.

If the ground impedance is considered locally reacting, such as grass covered ground, the total field includes also a ground wave term which arises from diffraction of waves from the surface. The total field is given by [Reference 5 as modified by Reference 6]:

$$p = \frac{\exp(ikR_1)}{R_1} + R_p \frac{\exp(ikR_2)}{R_2} + (1-R_p) F \frac{\exp(ikR_2)}{R_2} \tag{9}$$

where

$$F = 1 + \pi w e^{w^2} \operatorname{erfc}(-w) \tag{10}$$

and

$$w^2 = - \frac{i kR_2}{2 \cos^2\psi} (\beta + \sin\psi)^2 . \tag{11}$$

The additional term, which becomes significant when angle $\psi$ is small, is referred to as the Ground Wave. If angle $\psi$ is large, then the term F becomes negligible. For a bare ground or one covered by a thin layer of debris, the locally reacting surface impedance does not model the ground surface accurately. Instead, the ground surface is treated as the surface of a semi-infinite solid, which allows for wave penetration into the ground. A model for propagation near the interface of two semi-infinite media [7] is:

$$p = \frac{\exp(ik_a R_1)}{R_1} + R_p \frac{\exp(ik_a R_2)}{R_2} + BF\ (1-R_p) \frac{\exp(ik_a R_2)}{R_2} \quad (12)$$

where the factor B is a correction for two-media wave propagation given in [7], F is defined in Equation 10 with

$$w^2 = -\frac{ik_a R_2}{2}\left[\sin\psi + \beta_c \frac{1-1/n^2}{1-M^2}\right]^2 \quad (13)$$

where $\beta_c = \rho_a c_a / \rho_g c_g$,

$n = c_a/c_g$, $M = \rho_a/\rho_g$

with the indices a and g refer to air and ground, respectively. It should be emphasized that the influence of the ground wave term is significant only when both the source and the receiver are close to ground. If either source or receiver is located few wavelengths above the ground, the ground wave contribution is negligible. Furthermore, the ground wave ray has a phase term which is the same as the reflected ray, so that they can be added together to give a corrected reflection coefficient.

To take account of the ground when a barrier is present, one must account for all the reflected or reflected-diffracted rays. For a double-thin-wall barrier these reflected and diffracted rays are shown in Figure 10. This means that four paths are possible instead of the one shown in Figure 6. Thus, the calculations for the diffracted pressure in Equation 4 must be repeated four times, with different input and exit angles and radii. However, these calculations for the diffracted pressure at each frequency are added with proper phases, since they eminate from the same source. The reflected rays would include the reflection coefficient to modify their strengths. The influence of the ground wave is negligible since the receiver for the first reflection is located on top of the barrier, which usually measures few wavelengths in height.

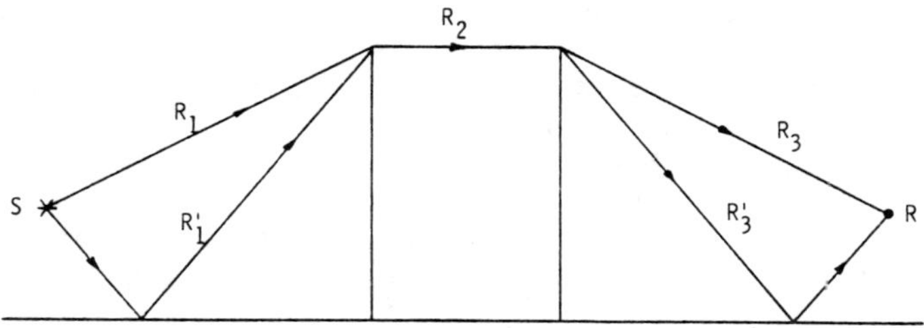


Figure 10. Geometry of reflected-diffracted rays.

The calculation of the noise level at the receiver before and after the erection of a barrier can be predicted as was discussed in this section. The ratio of the former to the latter gives the insertion loss. A typical contour plot of the insertion loss for the same traffic conditions of Figure 8 is shown in Figure 11 for a thin-wall-barrier over hard ground. It can be seen that reflection off the ground reduces the barrier noise reduction efficiency by 1-2dB for trucks and by 2-10dB for automobiles over the entire receiver field.

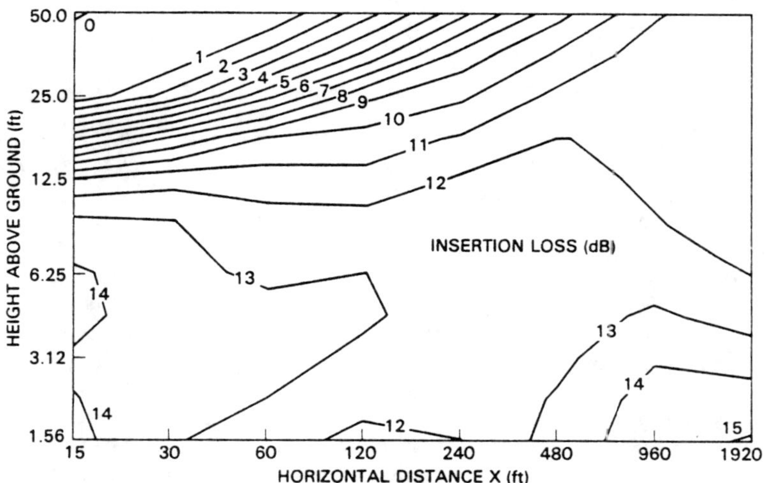


Figure 11. Contours of IL for a heavy truck line source at offset distance of 60 ft.

If the ground cover under the receiver and source fields is not uniform, such as hard pavement adjacent to an absorbent field of grass, then the uniform impedance models described above are not adequate. A new model which accounts for the discontinuity of the ground impedance was developed in Reference [8]. This model accounts for the diffraction of waves at the edge of an impedance discontinuity. Also, two separate ground waves are generated, which account for non-negligible levels when the source and receiver are near the ground. A comprehensive review and analysis of the literature on propagation over ground with corrections is given in Reference [9].

## SCALE MODEL TESTS

To verify the analytic model predictions for the barriers of Figure 3, tests were made on 1:5 scale models of the barriers in a large gymnasium on the University Park campus of the Pennsylvania State University. A carriage carrying eight speakers was commanded by computer to make forty (40) stops at intervals of two feet for a total half-length of scaled line of traffic of 80 ft. (24m). This would correspond to a traffic line of 800 ft. (240m). At each stop, one speaker was pulsed with two pure tone frequencies corresponding to two of the 19 (1/3-octave) band centered frequencies of the spectrum, and then another speaker is pulsed with a different set of two pure-tone frequencies. An array of sixteen microphones were used to record the sound field to distances up to 80 ft. (24m) away from the barrier. Each receiver pulse was recorded separately at each frequency at each stop of the cart. The computer was then used to process the data and perform the frequency and line source incoherent summations. A schematic diagram of the set-up is shown in Figure 12.

The first set of tests were conducted on hard floor, a floor totally covered with an outdoor carpet simulating grass cover for a uniform impedance coverage and finally tests on partially covered floor to simulate the impedance discontinuity of pavement-grassy terrain. The results of these tests are given in Reference [6].

The second phase of the testing program was the measurement of the insertion loss of the barriers in Figure 3. These barriers were made of 3/4" plywood and were tested either bare or covered by an absorbent carpet. The floor under the receiver field was either hardwood or covered by outdoor carpets. A summary of the results of these tests are given in Reference [4]. Typical experimental results for a double-thin-wall barrier are shown in Figure 13. Here the designation "primary" denotes the excess attenuation with only the reflections over the two sides of the barrier

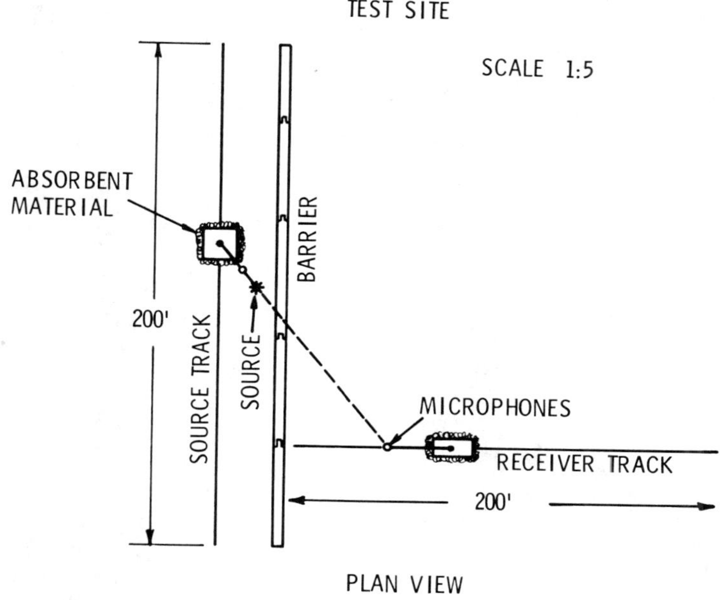


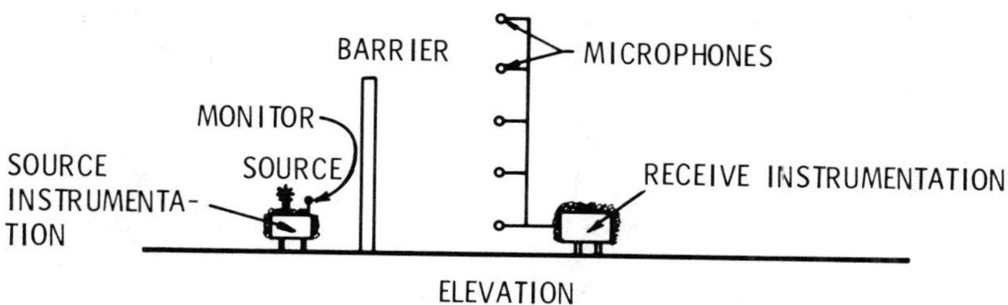


Figure 12. Schematic diagram for scale-model test setup.

included (4 paths). The "improved" version denotes the addition of multiple reflections and diffraction rays inside the space between the two thin walls for a total of 16 paths.

SUMMARY

Prediction methodology of highway noise before and after the erection of noise barriers was presented. The highway noise computer code allows for a variety of barrier shapes

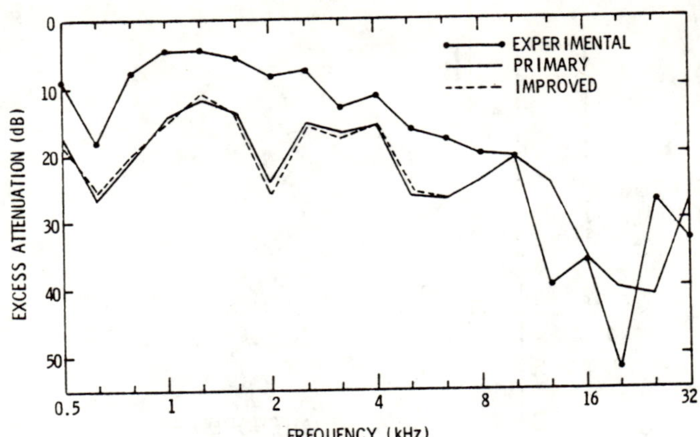


Figure 13. Theoretical and experimental data for a scale-model double-thin-wall barrier, height is 3 ft., for a line simulating heavy trucks at offset distance of 12 ft. and height of 1.6 ft. above hard ground. The receiver is located 68 ft. from the barrier and 3 in. above ground.

with singly or multiply reflected and diffracted rays. The presence of the ground is accounted for in the computation of the insertion loss of a noise barrier.

ACKNOWLEDGMENT

The author wishes to thank Drs. J. M. Lawther and R. P. Kendig for valuable discussions and contributions to the outcome of this project, Mr. M. Nobile for his help in the computational part of this work and Mr. D. Tate for making the results of the tests available. Special thanks goes to Mrs. Vickie Nagle for typing the manuscript and the Naval Ocean Systems Center, Code 635, for encouragement and support. This work was supported by the National Cooperative Highway Research Program, The Transportation Research Board, National Research Council.

## REFERENCES

1. Keller, J. B. "Geometrical Theory of Diffraction," *J. Opt. Soc. Am.* 52:116-130 (1962).

2. Readfern, S. W. "Some Acoustic Source-Observer Problems," *Phil. Mag.* 30:223-236 (1940).

3. Hayek, S. I., Lawther, J. M., Kendig, R. P. and Simowitz, K. T. "Investigation of Selected Noise Barrier Acoustical Parameters," Final Report, National Cooperative Highway Research Report, 3-26, April 1978.

4. Lawther, J. M., Hayek, S. I., Tate, D. C. and Nobile, M. A. "Theoretical and Experimental Investigations of Selected Noise Barrier Acoustical Parameters," Final Report, National Cooperative Highway Research Program, 3-26, February 1980.

5. Lawhead, R. B. and Rudnick, I. "Acoustic Wave Propagation Along a Constant Normal Impedance," *J. Acoust. Soc. Am.* 23:546-549 (1951).

6. Hayek, S. I., Lawther, J. M. and Tate, D. C. "Model for Acoustic Propagation over Ground - Theory and Experiment," National Cooperative Highway Research Program, Tech. Report, 3-26, May 1980.

7. Attenborough, K., Hayek, S. I. and Lawther, J. M. "Propagation of Sound Above a Porous Half-Space," *J. Acoust. Soc. Am.* 68:1493-1501 (1980).

8. Naghieh, M. and Hayek, S. I. "Diffraction of a Point Source by Two-Impedance Covered Half-Planes," *J. Acoust. Soc. Am.* 69:629-637 (1981).

9. Hayek, S. I., Attenborough, K. and Lawther, J. M. "Models for Acoustic Propagation over Absorbent Ground," The Federal Highway Administration, DOT-FH-11-9515, Interim Report, August 1980.

## BIBLIOGRAPHY ON GROUND EFFECTS

Alavi, H. and Fillippi, P. J. T. "On the resolvant of the Perkeris operator with a Dirichlet condition," *J. Sound Vib.* 237-243 (1978).

Attenborough, K. "Sound attenuation over ground cover," *Shock and Vibration Digest* 10(7):3-13 (1978).

Attenborough, K., Hayek, S. I. and Lawther, J. M. "Propagation of sound above a porous half-space," *J. Acoust. Soc. Am.* 68:1493-1501 (1980).

Banos, Jr., Alfredo and Wesley, J. P. "The horizontal electric dipole in a conducting half-space," Univ. of Calif. Marine Physical Laboratory, SIO Reference 53-33 (1953).

Banos, Jr., Alfredo and Wesley, J. P. "The horizontal electric dipole in a conducting half-space, Part II," Univ. of Calif. Marine Physical Laboratory, SIO Reference 54-31 (1954).

Banos, Jr., A. *Dipole Radiation in the Presence of a Conducting Half-Space*, (New York: Pergamon Press, 1966).

Bass, H. and Bolen, L. "Propagation of Sound Through the Atmosphere: Effects of Ground Cover," Univ. of Mississippi Physical Acoustics Research Group Report 78-01 (1978).

Brekhovskikh, L. M. "The reflection of spherical waves at a plane interface between two media," *J. Tech. Phys. U.S.S.R.* 18:455 (1948a).

Brekhovskikh, L. M. "The field of refracted electromagnetic waves in the problem of the point source," *Izv. Akad. Nauk. U.S.S.R. Ser. Fiz.* 12:322 (1948b).

Brekhovskikh, L. M. "The reflection and refraction of spherical waves," *Uspekhi Fiz. Nauk.* 38:1 (1949).

Brekhovskikh, L. M. *Waves in Layered Media*, (New York and London: Academic Press, 1960).

Briquet, M. and Filippi, P. J. T. "Diffraction of a spherical wave by an absorbing plane," *J. Acoust. Soc. Am.* 61(3): 640-646 (1977).

Chien, C. F. and Soroka, W. W. "Sound propagation along an impedance plane," *J. Sound Vib.* 43(1):9-20 (1975).

Clemmow, P. *The Plane Wave Spectrum Representation of Electromagnetic Fields*, (Pergamon Press, 1966).

Donato, R. J. "Propagation of a spherical wave used a plane boundary with a complex impedance," *J. Acoust. Soc. Am.* 60(1):34-39 (1976a).

Donato, R. J. "Spherical-wave reflection from a boundary of reactive impedance using a modification of Cagniard's method," *J. Acoust. Soc. Am.* 60(5):999-1002 (1976b).

Embleton, T. F. W., Piercy, J. E. and Olson, N. "Outdoor sound propagation over ground of finite impedance," *J. Acoust. Soc. Am.* 59(2):267-277 (1976).

Erdelyi, A. *Asymptotic Expansions,* (New York: Dover, 1956).

Ewing, W. M., Jardetsky, W. S. and Press, F. *Elastic Waves in Layered Media* (McGraw-Hill, 1957).

Felsen, L. B. and Marcuvitz, N. *Radiation and Scattering of Waves* (New Jersey: Prentice-Hall, Inc., 1973).

Fillippi, P. J. T. and Habault, D. "Reflexion of a spherical wave by the plane interface between a perfect fluid and a porous medium," *J. Sound Vib.* 56:97-103 (1978).

Graff, K. F. *Wave Motion in Elastic Solids* (Ohio State University Press, 1975).

Habault, D. and Filippi, P. J. T. "On the resolvant of the Perkeris operator with a Newmann condition," *J. Sound Vib.* 56:87-95 (1978).

Hayek, S. I., Lawther, J. M., Kendig, R. P. and Simowitz, K. T. "Investigation of selected noise barrier acoustical parameters," Final Report, National Cooperative Highway Research Program 3-26, Transportation Research Board, National Research Council,(1978).

Hayek, S. I., Lawther, J. M. and Tate, D. C. "Model for acoustical propagation over ground - theory and experiment," Technical Report for National Cooperative Highway Research Program 3-26, Transportation Research Board, National Research Council (1980).

Hayek, S. I., Attenborough, K. and Lawther, J. M. "Models for acoustic propagation over absorbent ground," The Federal Highway Administration, DOT-FH-11-9515, Interim Report (1980).

Heins, A. and Feshback, F. "On the coupling of two half-planes," In: *Proceedings of Symposia in Applied Mathematics* (McGraw-Hill, 1954) 5:75-87.

Ingard, U. "On the reflection of a spherical sound wave from an infinite plane," *J. Acoust. Soc. Am.* 23:329 (1951).

Isei, T., Embleton, T. F. W. and Piercy, J. E. "Noise reduction by barriers on finite impedance ground," *J. Acoust. Soc. Am.* 67:46-58 (1980).

Lawhead, R. B. and Rudnick, I. "Measurements on an acoustic wave propagated along a boundary," *J. Acoust. Soc. Am.* 23:541-545 (1951a).

Lawhead, R. B. and Rudnick, I. "Acoustic wave propagation along a constant normal impedance boundary," *J. Acoust. Soc. Am.* 23:546-549 (1951b).

Lawther, J. M., Hayek, S. I., Tate, D. C. and Nobile, M. A. "Theoretical and experimental investigations of selected noise barrier acoustical parameters," Final Report, National Cooperative Highway Research Program 3-26 (1980).

Lawther, J. M. and Hayek, S. I. "A noise barrier parameter study," *Proceedings of Conference on Highway Traffic Noise Mitigation,* Los Angeles, CA, Dec. 11-15, p. 172-184 (1978).

Naghieh, M. and Hayek, S. I. "Diffraction of a point source by two-impedance covered half-planes," *J. Acoust. Soc. Am.* 69:629-637 (1981).

Norton, K. A. "The propagation of radio waves over the surface of the earth in the upper atmosphere," *Proc. IRE* 24:1367-1387 (1936).

Norton, K. A. "The propagation of radio waves over the surface of the earth and in the upper atmosphere," *Proc. IRE* 25:1203-1236 (1937).

Pao, S., Wenzel, A. and Oncley, P. "Prediction of ground effects on aircraft noise," NASA Technical Paper 1104 (1978).

Paul, D. I. "Acoustical radiation from a point source in the presence of two media separated by a plane interface," Doctoral Dissertation, Univ. of Calif., Los Angeles (1956).

Paul, D. I. "Acoustical radiation from a point source in the presence of two media," *J. Acoust. Soc. Am.* 29:1102-1109 (1957).

Paul, D. I. "Wave propagation in acoustics using the saddle-point method," *J. Math. and Phys.* 38:1-15 (1959).

Rudnick, I. "The propagation of an acoustic wave along a boundary," *J. Acoust. Soc. Am.* 19:348-356 (1947).

Skudrzyk, E. J. *The Foundations of Acoustics*, (Vienna: Springer, 1971).

Sommerfeld, A. "Uber die Ausbretiung der Wellen in der drahtlosen Telegraphie," *Ann. Physik* 28:665-737 (1909).

Sommerfeld, A. *Partial Differential Equations in Physics* (New York and London: Academic Press, 1949).

Stickler, D. C. "Reflected and lateral waves for the Sommerfeld model," *J. Acoust. Soc. Am.* 60(5):1061-1070 (1976).

Stratton, J. A. *Electromagnetic Theory* (New York: McGraw-Hill, 1941).

Thomasson, S. I. "Reflection of waves from a point source by an impedance boundary," *J. Acoust. Soc. Am.* 59(4): 780-785 (1976).

Thomasson, S. I. "Sound propagation above a layer with a large refractive index," *J. Acoust. Soc. Am.* 61(3):659-674 (1977a).

Thomasson, S. I. "Theory and experiments on sound propagation above an impedance boundary," Report No. 75, Department of Building Acoustics, Lund Institute of Technology, Lund, Sweden (1977b).

Van Der Pol, B. "Theory of the reflection of light from a point source by a finitely conducting glat mirrow, with an application to radio-telegraphy," *Physica* 2:843-854 (1935).

Van Der Waerden, B. L. "On the method of saddle-points," *Appl. Sci. Research* B2:33-46 (1951).

Wait, J. R. *Electromagnetic Waves in Stratified Media* (Oxford: Pergamon Press, 1962).

Watson, G. N. *A Treatise in the Theory of Bessel Functions* (Cambridge: University Press, 1944).

Wenzel, A. R. "Propagation of waves along an impedance boundary," *J. Acoust. Soc. Amer.* 55:956-963 (1974).

Weyl, H. "Ausbreitung elektromagnetischer Wellen uber einem ebenen Leitner," *Ann. Physik* 60:481-500 (1919).

Weyl, H. "Erwiderung auf Hrn. Sommerfelds Bemerkungen uber die Ausbreitung der Wellen in der drahtlosen Telegraphie," *Ann. Physik* 62:482-485 (1920).

BIBLIOGRAPHY ON BARRIER AND DIFFRACTION

Ahluwalia, D. S., Lewis, R. M. and Boersma, J. "Uniform asymptotic theory of diffraction by a plane screen," *SIAM J. Appl. Math.* 16:783-807 (1968).

Ahluwalia, D. S. "Uniform asymptotic theory on diffraction by the edge of a three dimensional body," *SIAM J. Appl. Math.* 18:287-301 (1970).

Bowman, J. J., Senior, T. B. A. and Uslenghi, P. L. E. *Electromagnetic and Acoustic Scattering by Simple Shapes,* (Amsterdam, The Netherlands: North-Holland Pub., 1969).

Bremmer, H. *Terrestrial Radio Waves* (New York: Elsvier Pub. Co., 1949).

Bromwich, T. J. "Diffraction of waves by a wedge," *Proc. Lond. Math. Soc.* 14:450-463 (1915).

Butler, G. F. "A note on improving the attenuation given by a noise barrier," *J. Sound Vib.* 32(3):367-369 (1974).

Carslaw, H. S. "Diffraction of waves by a wedge of any angle," *Proc. Lond. Math. Soc.* 18:291-306 (1920).

Carslaw, H. S. "Some multiform solutions of the partial differential equations of physics and mathematics and their applications," *Proc. Lond. Math. Soc.* 30:121-163 (1899).

Clemmow, P. C. *The Plane Wave Spectrum of Electromagnetic Waves* (London: Pergammon Press, 1966).

Embleton, T. W. F. "Line integral theory of barrier attenuation," *J. Acoust. Soc. Am.* 67:42-45 (1980).

Franz, W. and Klante, K. "Diffraction by surfaces of variable curvature," *IRE Trans. on Antennas and Propagation* AP-7: S68-S70 (1959).

Fujiwara, F., Ando, Y. and Maekawa, Z. "Attenuation of a spherical sound wave diffracted by a thick plate," *Acustica* 28(6):341-347 (1973).

Gordon, C. G., et al. "Highway noise - a design guide for highway engineers," *NCHRP Report No. 117* (1971).

Hayek, S. I. "Acoustic diffraction by impedance edges," *J. Acoust. Soc. Am.* 58:(Supplement 1):S129 (1975).

Hayek, S. I., Lawther, J. M., Kendig, R. P. and Simowitz, K. T. "Investigation of selected noise barrier acoustical parameters," Final Report National Cooperative Highway Research Program 3-26, Transportation Research Board, National Research Council (1978).

Hayek, S. I., Lawther, J. M. and Kendig, R. P. "Effectiveness of Cylindrical Topped Noise Barriers," *J. Acoust. Soc. Am.* 65(Supplement 1):S65 (1979).

Hayek, S. I. and Nobile, M. A. "Diffraction by absorbent wide barriers," *J. Acoust. Soc. Am.* 69(Supplement 1): S101 (1981).

Hong, S. "Asymptotic theory of electromagnetic and acoustic diffraction by smooth convex surfaces of variable curvature," *J. Math. Phys.* 8:1223-1232 (1967).

Hutchins, D. L. and Kouyoumijan, R. G. "Calculation of the field of a baffled array by the geometrical theory of diffraction," *J. Acoust. Soc. Am.* 45:485-492 (1969).

Isei, T., Embleton, T. F. W. and Piercy, J. E. "Noise reduction by barriers on finite impedance ground," *J. Acoust. Soc. Am.* 67:46-58 (1980).

Jonasson, H. G. "Diffraction by wedges of finite acoustic impedance with application to depressed roads," *J. Sound Vib.* 25:577-585 (1972).

Jonasson, H. G. "Sound reduction by barriers on the ground," *J. Sound Vib.* 22:113-126 (1972).

Jones, D. S. "Diffraction by a thick semi-infinite plate," *Proc. R. Soc.* A217:153-175 (1953).

Keller, J. B. "A geometrical theory of diffraction," In: *Calculus of Variations and Its Applications*, L. M. Graves, Ed. (New York: McGraw-Hill, 1958), pp. 27-52.

Keller, J. B. "Diffraction by a convex cylinder," *IRE Trans. on Antennas and Propagation* AP-4:312-321 (1956).

Keller, J. B. "Diffraction by an aperture," *J. Appl. Phys.* 28:426-444 (1957).

Keller, J. B. "Geometrical theory of diffraction," *J. Opt. Soc. Am.* 52:116-130 (1962).

Keller, J. B. and Blank, A. "Diffraction and reflection of pulses by wedges and corners," *Commun. Pure Appl. Math.* 4:75-95 (1951).

Keller, J. B. and Levy, B. R. "Decay exponents and diffraction coefficients for surface waves on surfaces of nonconstant curvature," *IRE Trans. on Antennas and Propagation* AP-7:S52-S61 (1959).

Kendig, R. P. *Acoustic Diffraction by an Impedance Covered Half-Plane,* Ph.D. Thesis, The Pennsylvania State Univ., University Park, PA (1977).

Kouyoumjian, R. G. and Pathak, P. H. "A uniform geometrical theory for an edge in a perfectly conducting surface," *Proc., IEEE* 62(11):1448-1461 (1974).

Kugler, B. A., Commins, D. E. and Galloway, W. J. "Highway noise - a design guide for prediction and control," *NCHRP Report No. 174* (1976).

Kurze, U. and Anderson, G. S. "Sound attenuation by barriers," *J. Appl. Acoust.* 4(1):35-53 (1971).

Kurze, U. J. "Noise reduction by barriers," *J. Acoust. Soc. Am.* 55:504-518 (1974).

Lawther, J. M., Hayek, S. I., Tate, D. C. and Nobile, N. A. "Theoretical and experimental investigations of selected noise barrier acoustical parameters," Final Report, National Cooperative Highway Research Program 3-26 (1980).

Lawther, J. M. and Hayek, S. I. "A noise barrier parameter study," *Proceedings of Conference on Highway Traffic Noise Mitigation,* Los Angeles, CA, Dec. 11-15, pp. 172-184 (1978).

Levy, B. R. and Keller, J. B. "Diffraction by a smooth object," *Commun. Pure and Appl. Math.* 12:159-209 (1959).

Ludwig, D. "Uniform asymptotic expansion of the field scattered by a convex object at high frequencies," *Commun. Pure Appl. Math.* 20:103-138 (1967).

MacDonald, H. M. "A class of diffraction problems," *Proc. Lond. Math. Soc.* 14:410-427 (1915).

Maekawa, Z. "Noise reduction by screen of finite size," *Mem. Fac. Eng.,* Kobe Univ. 12:1-12 (1966).

Maekawa, Z., Fujiwara, K. and Morimoto, M. "Some problems of noise reduction by barriers," *Symp. on Noise Prevention,* Miskoic, Hungary, paper 4.8 (1971).

Maekawa, Z. "Noise reduction by screens," *Appl. Acoust.* 1:157-173 (1968).

Malyuzhinets, G. D. "Sound radiation by vibration sides of an arbitrary wedge," *Akustichesky Zhurnal* 1:144-163 (1955).

Malyuzhinets, G. D. "The radiation of sound by the vibrating boundaries of an arbitrary wedge, Part I," *Sov. Phys. Acoust.* 1:152-174 (1955).

Malyuzhinets, G. D. "Radiation of sound from vibrating faces of an arbitrary wedge, Part II," *Sov. Phys. Acoust.* 1:240-248 (1955).

Malyuzhinets, G. D. "Conversion formula for Sommerfeld integral," *Doklady Akademii Nauk S.S.S.R.* 118:1099-1102 (1958).

Malyuzhinets, G. D. "The excitation, reflection, and emission of surface waves on a wedge with given impedances of the faces," *Doklady Akademii Nauk S.S.S.R.* 121:436-439 (1958).

Nobile, M. A., Hayek, S. I. and Lawther, J. M. "A comprehensive model for the straight edge noise barrier," *J. Acoust. Soc. Am.* 68(Supplement 1):S54 (1980).

Nobile, M. A. and Hayek, S. I. "A new model for a noise barrier on a rigid ground plane," *J. Acoust. Soc. Am.* 69(Supplement 1):S102 (1981).

Oberhettinger, F. "Diffraction of waves by a wedge," *Commun. Pure Appl. Math.* 7:551-564 (1954).

Pathak, P. H. and Kouyoumjian, R. G. "The dyadic diffraction coefficient for a perfectly-conducting wedge," Electro Science Laboratory, Dept. Elec. Eng., Report 2183-4, Ohio State Univ., Columbus, Ohio, AD 707 827 (1970).

Pathak, P. H. and Kouyoumjian, R. G. "An analysis of the radiation from apertures in curved surfaces by the geometrical theory of diffraction," *Proc. of the IEEE* 62: 1438-1447 (1974).

Pierce, A. D. "Diffraction of sound waves around corners and over wide barriers," *J. Acoust. Soc. Am.* 55:941-955 (1974).

Pierce, A. D. and Hadden, W. J., Jr. "Noise diffraction around barriers of finite acoustic impedance," *Proc. 3rd Interagency Symposium on University Research in Transportation Noise,* U.S. Dept. of Transportation, 50-58 (1975).

Pierce, A. D. and Hadden, W. J., Jr. "Plane wave diffraction by a wedge with finite impedances," *J. Acoust. Soc. Am.* 63:17-27 (1978).

Redfern, S. W. "Some acoustic source-observer problems," *Phil. Mag.* 30:223-236 (1940).

Rawlins, A. D. "The solution of a mixed boundary value problem in the theory of diffraction by a semi-infinite plane," *Proc. Roy. Soc. Lond.* A346:469-484 (1974).

Ross, R. A. and Hamid, M. A. K. "Scattering by a wedge with rounded edge," *IEEE Trans. on Antennas and Propagation* AP-19(4):507-515 (1971).

Sakharova, M. P. "Influence of the edge of a wedge with vibrating faces on the radiated acoustic power," *Sov. Phys. - Acoustics* 12 (1966).

Scholes, W. E., Salvidge, A. C. and Sargent, J. W. "Field performance of a noise barrier," *J. Sound Vib.* 16:627-642 (1971).

Senior, T. B. A. "Diffraction by a semi-infinite metallic sheet," *Proc. Roy. Soc. Lond.* A213:436-458 (1956).

Simpson, M. A. "Noise barrier design handbook," FHWA Report No. RD-76-58 (1976).

Skudrzyk, E. J. "The exact theory of impedance covered straight edge and wedge," Applied Research Laboratory Report No. 75-11, The Pennsylvania State Univ., University Park, PA (1975).

Skudrzyk, E. J. *The Foundations of Acoustics* (Vienna: Springer, 1971).

Sommerfeld, A. "Mathematische Theorie der diffraction," *Math. Ann.* 47:317-374 (1896).

Tuzhilin, A. A. "New representations of diffraction fields in wedge-shaped regions with ideal boundaries," *Sov. Phys. - Acoustics* 9 (1963).

Voltmer, D. R. *Diffraction by Doubly Curved Convex Surfaces* Ph.D. Dissertation, The Ohio State University, Columbus, Ohio (1970).

Williams, W. E. "Diffraction by an imperfectly conducting right-angled wedge," *Proc. Camb. Philos. Soc.* 55:195-209 (1959).

Williams, W. E. "Diffraction of an E-polarized plane wave by an imperfectly conducting wedge," *Proc. Roy. Soc. Lond.* A-252:376-393 (1959).

Williams, W. E. "Propagation of electromagnetic surface waves along wedge surfaces," *Quart. J. Mech. Appl. Mat.* 13:277-284 (1960).

CHAPTER 4

# INVERSION OF FIRST ARRIVAL CRACK-SCATTERING DATA

*A. Norris and J. D. Achenbach*
The Technological Institute
Northwestern University
Evanston, IL 60201

## ABSTRACT

The problem of locating cracks and determining their extent is of fundamental concern in the non-destructive evaluation of materials. In this paper we consider the inverse problem of determining the geometrical characteristics of a crack from the diffraction of an incident wave by the crack-edge. It is assumed that the material is homogeneous and isotropic. By the use of elastodynamic ray theory we have obtained an inversion procedure which uses only the time delay of the first received diffracted signal as input. For a set of observations the method yields the position of a "flashpoint" on the crack edge. Several flashpoints constitute a segment of the crack edge from which we can infer the plane of the crack. Each observation in the time domain contains information on a second flashpoint on the crack edge. The time delay for this flashpoint can be obtained from the periodicity of the high-frequency amplitude spectrum. The inversion procedure is then again applied to yield another segment of the crack edge. In general, it may be expected that the two segments will adequately define the size of the crack.

## INTRODUCTION

We consider the problem of inversion of crack-scattering data. The configuration is as follows: A source S produces a signal which is scattered by a flat crack with a convex edge. The scattered signal is received by an observer Q, at a position different from S. Both S and Q can be moved, possibly within the limits of certain constraints. The problem to be solved is: Given the received signal at Q and the freedom to position S and Q, develop a procedure to determine a sufficient number of points on the crack-edge.

As a preliminary to formulating the inverse method, it is necessary to consider the direct problem. In the time domain, the first arriving signals at Q essentially come from two flashpoints on the crack edge. By well known results of Fourier analysis, this effect displays itself in the frequency domain as regular oscillations of the amplitude at high frequency. Thus, for a given source and observer, the direct problem indicates that the received signal contains two informative quantities. The first is the arrival time of the first received signal, from which we determine the travel time corresponding to the position of one of the flashpoints. The second is the travel time corresponding to the other flashpoint, which may be inferred by measuring the period of the oscillations in the high frequency spectrum.

An inverse method using just these two travel times is described. In order to apply the method, it is necessary to take measurements for different positions of S and Q. By taking enough measurements it is possible to map out the crack edge as accurately as is desired.

The inverse method presented here is different from that of [1]. The theory of Reference 1 is based on the physical optics or Kirchhoff approximation, and is computationally quite arduous. The present theory follows from Fermat's Principle and the asymptotically exact geometrical theory of diffraction. Also, its application is straightforward as demonstrated by example.

In section 2 we describe the geometrical theory of diffraction. This theory is used in section 3 to consider the direct problem for a flat crack with a convex edge. The inverse method is described in section 4, and as an example, the method is applied to synthetic data for an elliptical crack.

## GEOMETRICAL THEORY OF DIFFRACTION

The geometrical theory of diffraction (GTD) for linear elastodynamics is based on the result that two cones of diffracted rays are generated when a ray carrying a high-frequency longitudinal wave strikes the edge of a crack. The inner and outer cones consist of rays of longitudinal and transverse motion, respectively. For cracks in elastic solids the three-dimensional theory of edge-diffraction was discussed by Achenbach and Gauteson [2].

Figure 1 shows an incident ray of longitudinal motion and the corresponding cones of diffracted rays. The angle of the incident ray with the edge is $\phi_L$. Thus

$$\cos\phi_L = \underset{\sim}{p} \cdot \underset{\sim}{t} \tag{1}$$

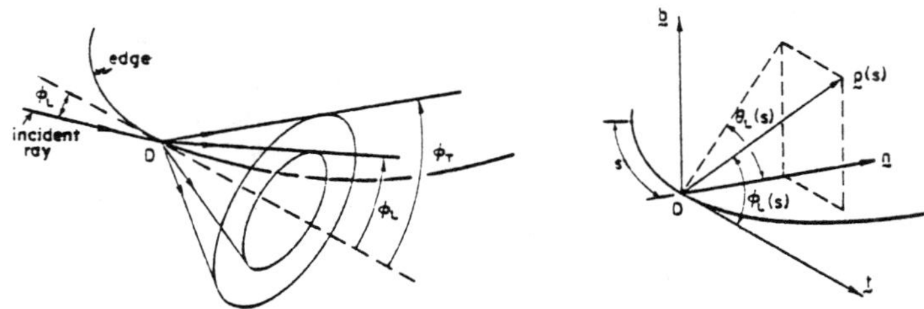


Figure 1. Cones of diffracted longitudinal and transverse rays for an incident longitudinal ray.

where $\underset{\sim}{p}$ defines the direction of propagation along the incident ray and $\underset{\sim}{t}$ is a unit vector along the tangent to the edge, chosen in the direction which makes $\phi_L$ acute. The half-angles $\phi_\beta$ of the cones of diffracted rays of type $\beta(\beta=L,T)$ are given by $\phi_L$ and $\phi_T$, where

$$\cos\phi_T = (c_T/c_L)\cos\phi_L \tag{2}$$

For time harmonic motion and $k_\beta R_\beta >> 1$, the field on the diffracted ray of type $\beta = L,T$ may be expressed in the form

$$\underset{\sim}{u}^L = U_o[k_\beta R_\beta(1+R_\beta/\rho_\beta)]^{-1/2} D_\beta^L(\theta;\phi_L,\theta_L)\exp(ik_\beta R_\beta)\underline{d}^\beta \tag{3}$$

where $U_o$ defines the amplitude and phase on the incident ray at the point of diffraction, $D_\beta^L(\theta;\phi_L,\theta_L)$ is the diffraction coefficient, (the angle $\theta_L$ is defined in Figure 1), $\underline{d}^\beta$ is the unit vector which defines the displacement direction, $R_\beta$ is the distance along a diffracted ray measured from the point and $\rho_\beta$ is the distance from the point of diffraction to the other caustic. An explicit expression for $\rho_\beta$ is

$$\rho_\beta = -a\sin\phi_\beta[ad\phi_\beta/ds + \cos\theta]^{-1} \tag{4}$$

where s is arc length measured along the edge and a is the signed radius of curvature of the edge.

The diffraction coefficient $D_\beta^L(\theta;\phi_L,\theta_L)$ can be obtained by solving the canonical problem of diffraction of plane waves by a semi-infinite crack. For crack-faces which are free of surface tractions (an empty crack) this was done by Achenbach and Gautesen [2].

In this paper we consider only longitudinal to longitudinal diffraction. The diffraction coefficient $D_L^L(\theta;\phi_L,\theta_L)$ follows from [2] as

$$D_L^L(\theta;\phi_L,\theta_L) = e^{i\pi/4}\,(f_1 + f_2)/f_3 \tag{5}$$

where

$$f_1 = [\kappa_T^2 - 2(\eta^2 + \rho^2)][\kappa_T^2 - 2(\eta^2 + \xi^2)]\sin\left(\frac{1}{2}\theta_L\right)\cos\left(\frac{1}{2}\theta\right)$$

$$f_2 = 2\sin\phi_L\sin\theta_L\sin\theta(\kappa_T\sin\phi_T - \rho)^{1/2}(\kappa_T\sin\phi_T - \xi)^{1/2}[\eta^2 - \rho\xi$$

$$+ \eta(\rho + \xi)\tan(\phi_T - \phi_R + \Gamma)]$$

$$f_3 = (2\pi)^{1/2}(1 - \kappa_T^2)(\kappa_R\sin\phi_R + \rho)(\kappa_R\sin\phi_R +$$

$$+ \xi)(\rho + \xi)\kappa^+(\rho)\kappa^+(\xi)$$

and

$$\eta = \cos\phi_L$$

$$\rho = \sin\phi_L\cos\theta_L$$

$$\xi = -\sin\phi_L\cos\theta$$

$$\kappa_\alpha = c_\alpha/c_L \quad , \quad \alpha = T,R$$

$$\cos\phi_\alpha = \kappa_\alpha^{-1}\cos\phi_L \quad , \quad \alpha = T,R$$

$$\exp(i\Gamma) = \kappa^+(i\eta)/\kappa^+(0)$$

$$\kappa^+(\lambda) = \exp\left[-\frac{1}{2\pi}\int_1^{K_T^2}\arctan\left(\frac{t(t-1)^{1/2}(\kappa_T^2-t)^{1/2}}{(t-(1/2)\kappa_T^2)^2}\right)\times\right.$$

$$\left.\times\frac{dt}{t - \eta^2 + \lambda(t - \eta^2)^{1/2}}\right]$$

The elastodynamic diffraction coefficients $D_\beta^\alpha(\theta;\phi_\alpha,\theta_\alpha)$, $\alpha,\beta$ = L,TV,TH are in general complex numbers whose phase depend upon $\theta$, $\phi_\alpha$, and $\theta_\alpha$. In the special case of $\alpha = \beta = L$, we have the simplification that $\exp(-i\pi/4)D_L^L(\theta;\phi_L,\theta_L)$ is a real number. We shall avail of this result later.

## L-L DIFFRACTION BY A FLAT CONVEX CRACK

Consider a flat crack with a smooth convex edge situated in a homogeneous, isotropic, linearly elastic solid. The crack is of zero thickness and its faces are traction-free. A source at S emits a signal. Consider an observer Q. In the time domain, the first arriving signals will be the L-L diffracted signals emanating from the two flashpoints on the crack edge which have cones of diffracted L-rays passing through the point Q. Using the diffraction theory of section 2, we now estimate this diffracted field at Q in the frequency domain.

A rectangular coordinate system $(x_1,x_2,x_3)$ is defined with origin inside the crack and such that the normal to the crack faces is in the $x_3$ direction. Since the crack edge is a smooth convex curve, we may parameterize it by an angle $\psi$, $0 \leq \psi < 2\pi$. Thus the crack edge is C, where

$$C = \{\underline{x}(\psi) \mid \underline{x}(\psi) = \ell(\psi)(\cos\psi,\ \sin\psi,\ 0),\ 0 \leq \psi < 2\pi\} \tag{6}$$

for some smooth $2\pi$-periodic function $\ell(\psi) > 0$. Following [3], at each point $\underline{x}(\psi) \in C$ we assign a right handed triad of unit vectors $\{\underline{n}(\psi),\ \underline{b}(\psi),\ \underline{t}(\psi)\}$, defined as

$$\underline{n}(\psi) = (-\cos\chi, -\sin\chi, 0) \tag{7a}$$

$$\underline{b}(\psi) = (0,0,-1) \tag{7b}$$

$$\underline{t}(\psi) = (\sin\chi, -\cos\chi, 0) \tag{7c}$$

where

$$\chi(\psi) = \psi - \arctan(\ell^{-1}(\psi)\ d\ell(\psi)/d\psi). \tag{8}$$

So, $\underline{n}(\psi)$ is the inwards normal to the edge, and $\underline{t}(\psi)$ is the tangent to the edge at $\underline{x}(\psi)$. We assume for simplicity that C is strictly convex, implying that the function $\chi(\psi)$ is invertible.

Let $\underline{x}_S$, $\underline{x}_Q$ define the points S and Q, and let $\psi_\alpha$, $\alpha$ = 1,2 be the angles of the two flashpoints for L-L diffraction. Define the distances $R_{i\alpha}$, $R_{L\alpha}$, $\alpha = 1,2$ as

$$R_{i\alpha} \equiv |\underline{x}_S - \underline{x}(\psi_\alpha)| \tag{9a}$$

$$R_{L\alpha} \equiv |\underline{x}_Q - \underline{x}(\psi_\alpha)|. \tag{9b}$$

Then the flashpoints are determined by the equation

$$\underline{t}(\psi_\alpha)\cdot[R_{i\alpha}(\underline{x}(\psi_\alpha) - \underline{x}_Q) + R_{L\alpha}(\underline{x}(\psi_\alpha) - \underline{x}_S)] = 0, \alpha=1,2. \quad (10)$$

In general Equation 10 has two distinct roots except in certain special circumstances such as when $\underline{x}_S//\underline{x}_Q//\underline{i}_3$, in which case the number of roots is infinite and each point on the edge is a flashpoint. We disregard such a possibility but note that it may be treated by suitably modifying the diffraction theory, see [3]. At either flashpoint, the diffraction angles $\phi_{L\alpha}$, $\theta_{L\alpha}$, and $\theta_\alpha$, $\alpha = 1,2$ follow from $0 \leq \phi_{L\alpha}$, $\theta_{L\alpha} < \pi$, $0 \leq \theta_\alpha < 2\pi$, where

$$\cos\phi_{L\alpha} = \underline{t}(\psi_\alpha)\cdot[\underline{x}(\psi_\alpha) - \underline{x}_S]/R_{i\alpha} \quad (11a)$$

$$\cos\theta_{L\alpha} = \underline{n}(\psi_\alpha)\cdot[\underline{x}(\psi_\alpha) - \underline{x}_S]/(R_{i\alpha}\sin\phi_{L\alpha}) \quad (11b)$$

$$\cos\theta_\alpha = \underline{n}(\psi_\alpha)\cdot[\underline{x}_Q - \underline{x}(\psi_\alpha)]/(R_{L\alpha}\sin\phi_{L\alpha}) \quad (11c)$$

$$\mathrm{sgn}(\sin\theta_\alpha) = -\mathrm{sgn}(x_{S3}\ x_{Q3}) . \quad (11d)$$

The distances from the flashpoints to their respective caustics are determined by Equations 4 and 11a, as $\rho_{L\alpha}$, $\alpha = 1,2$, where

$$\rho_{L\alpha}{}^{-1} = R_{i\alpha}{}^{-1} + (\cos\theta_{L\alpha} - \cos\theta_\alpha)/[a(\psi_\alpha)\sin\phi_{L\alpha}] \quad (12)$$

and $a(\psi) > 0$ is the radius of curvature at $\underline{x}(\psi)$, thus

$$a(\psi) = \ell(\psi)\ \sec(\psi - \chi)\ d\psi/d\chi . \quad (13)$$

Now consider the signal from S. Suppose it starts at $t = 0$. Then according to Huyghen's Principle, it is of the form

$$\underline{u}^i(\underline{x},t) = \underline{U}^i(\underline{x},t)\ H[c_L t - |\underline{x} - \underline{x}_S|] \quad (14)$$

where $H(\cdot)$ is the Heaviside step function and

$$\underline{U}^i(\underline{x},|\underline{x} - \underline{x}_S|/c_L) = U^i(\underline{x})(\underline{x} - \underline{x}_S)/|\underline{x} - \underline{x}_S| . \quad (15)$$

In the frequency domain,

$$\underline{u}^i(\underline{x},t) = (2\pi)^{-1} \int_{-\infty}^{\infty} \overline{\underline{u}}^i(\underline{x},\omega)\ e^{-i\omega t}\ d\omega \quad (16)$$

where

$$\overline{\underline{u}}^i(\underline{x},\omega) = \int_{|\underline{x}-\underline{x}_S|/c_L}^{\infty} \underline{U}^i(\underline{x},t)\ e^{i\omega t}\ dt . \quad (17)$$

In the high frequency regime we have

$$\underline{\bar{u}}^i(\underline{x},\omega) \sim \frac{i}{\omega} \underline{U}^i(\underline{x}, |\underline{x} - \underline{x}_S|/c_L)e^{ik_L|\underline{x}-\underline{x}_S|} \tag{18}$$

As the signal reaches the crack, each Fourier component diffracts according to the theory of section 2 from the two flashpoints $\underline{x}(\psi_\alpha)$, $\alpha = 1,2$. In the high frequency range, we have $k_L R_{L\alpha} >> 1$, $\alpha = 1,2$. Define the travel distances $R_\alpha$, $\alpha = 1,2$ as

$$R_\alpha = R_{i\alpha} + R_{L\alpha} \ . \tag{19}$$

Thus, in the frequency domain, the diffracted field at Q becomes, at high frequency,

$$\underline{\bar{u}}^d(\underline{x}_Q,\omega) \sim \omega^{-3/2} \sum_{\alpha=1,2} \underline{A}_\alpha e^{ik_L R_\alpha} \tag{20}$$

where the frequency independent vectors $\underline{A}_\alpha$, $\alpha = 1,2$ are

$$\underline{A}_\alpha = U^i(\underline{x}(\psi_\alpha))\ i(c_L/R_{L\alpha})^{1/2}(1 + R_{L\alpha}/\rho_{L\alpha})^{-1/2}D_L^L(\theta_\alpha;\phi_{L\alpha},\theta_{L\alpha}) \times$$

$$\times\ [\underline{x}_Q - \underline{x}(\psi_\alpha)]/R_{L\alpha} \ . \tag{21}$$

In practice, the observed signal is deconvoluted with respect to some response function, which, by Equation 18 is proportional to $\omega^{-1}$ at high frequency. Therefore, the observed frequency spectrum $V(\omega)$, follows from Equation 20 as

$$V(\omega) \sim \omega^{-1/2}|\underline{A}_1\ e^{ik_L R_1} + \underline{A}_2\ e^{ik_L R_2}| \tag{22}$$

at high frequency. Apart from the decrease proportional to $\omega^{-1/2}$, $V(\omega)$ oscillates with period $2\pi c_L/|R_2 - R_1|$. Also, from the definition of $\underline{A}_\alpha$, $\alpha$=1,2 and the result of section 2, that $e^{-i\pi/4}D_L^L(.;.,.)$ is real, we see that the relative maxima and minima of $V(\omega)$ at high frequency occur when $(\omega/c_L)|R_2-R_1|$ is an integral multiple of $\pi/2$.

Example: The Ellipse

Suppose the crack is an ellipse with major and minor semi-axes $\ell_M, \ell_m$. Define the aspect ratio $\beta \geq 1$ as $\beta = \ell_M/\ell_m$. Then all the preceding results follow for $\ell(\psi)$, $\chi(\psi)$ and $a(\psi)$ as

$$\chi(\psi) = \arctan(\beta^2\tan\psi) \tag{23}$$

$$\ell(\psi) = \ell_M[1 + (\beta^2 - 1)\sin^2\psi]^{-1/2} \tag{24}$$

$$a(\psi) = (\beta \ell_M)^{-2} [\ell(\psi) \cos\psi \sec\chi]^3 . \tag{25}$$

## INVERSE PROBLEM

From the results of the previous section, it is apparent that $R_1$ and $R_2$ are measurable quantities. Assuming $R_1 < R_2$, then $R_1$ is equal to $c_L \tau_o$, where $\tau_o$ is the elapsed time delay between emission of the signal at S and the arrival (at Q) of the first diffracted signal from the crack edge. The length $R_2$ may be determined by Fourier analyzing the received signal and measuring the period of the oscillations in the high frequency spectrum. If $\Delta\nu$ is the period in cycles per unit of time, then $R_2$ is given by

$$R_2 = R_1 + c_L/\Delta\nu . \tag{26}$$

In addition, as a check on the accuracy of $\Delta\nu$, we may use the result of the previous section that the relative maxima and minima of the spectrum can occur only at certain discrete frequencies. If $\nu$ is the frequency in cycles per unit time at one of these stationary points, then the result implies that $(4\nu/\Delta\nu)$ is a positive integer.

In the inverse problem, the position vectors $\underline{x}_S$ and $\underline{x}_Q$ and the travel distances $R_1$ and $R_2$ are known. The inverse problem is to deduce the two flashpoints on the crack edge.

Let R be either of $R_1$ or $R_2$ and let $\underline{x}_D$ be the associated flashpoint. On the basis of ray theory it then follows that $\underline{x}_D$ is located on an ellipsoid $E(\underline{x})$, whose foci are at the points $\underline{x}_S$ and $\underline{x}_Q$, and whose major-axis length is R. The eccentricity of the ellipsoid is

$$e = |\underline{x}_Q - \underline{x}_S|/R = |\underline{X}|/R \tag{27}$$

where we have implied

$$\underline{X} = \underline{x}_Q - \underline{x}_S . \tag{28}$$

The length of the minor axis is $(1 - e^2)^{1/2}R$, and the semi latus rectum, $\ell$, is

$$\ell = \frac{1}{2}(1 - e^2)R \tag{29}$$

For any point on the ellipsoid we have

$$\underline{p} = (\underline{x} - \underline{x}_S)/R_i \quad , \quad R_i = |\underline{x} - \underline{x}_S| \tag{30a,b}$$

We also define

$$\underline{q} = (\underline{x} - \underline{x}_Q)/R_L \quad , \quad R_L = |\underline{x} - \underline{x}_Q| . \tag{31a,b}$$

By using the equation for an ellipse in polar coordinates, centered at $\underline{x}$, $E(\underline{x})$ may be represented by

$$E(\underline{x}) = \{\underline{x}\,|\,\underline{x} = \underline{x}_S + R_i\underline{p};\ R_i = \ell[1 - e\ \underline{p}\cdot\underline{X}/|X|]^{-1},\ |\underline{p}| = 1 \tag{32}$$

The length $R_i$ can also be expressed as

$$R_i = W/2(R - \underline{p}\cdot\underline{X})\ ,\ \text{where } W = R^2 - |\underline{X}|^2 \tag{33}$$

To determine points on the crack edge we need additional source points and additional points of observation. Let us first move point Q over a small distance in the direction defined by the unit vector $\underline{V}$. As $\underline{x}_Q$ changes, so does the travel distance R and the ellipsoid $E(\underline{x})$. The distance $R_i$ for a point which is on the intersection of the two ellipsoids satisfies the equation

$$\nabla_{\underline{V}} R_i \equiv \underline{V}\cdot\underline{\nabla} R_i = 0\ . \tag{34}$$

By the use of Equation 34 we then find

$$\underline{p}\cdot\nabla_{\underline{V}}(\underline{X}/W) - \nabla_{\underline{V}}(R/W) = 0 \tag{35}$$

when W is defined by Equation 33. Similarly, if the source is moved over a small distance in the direction defined by the unit vector $\underline{U}$, we find

$$\underline{q}\cdot\nabla_{\underline{U}}(\underline{X}/W) + \nabla_{\underline{U}}(R/W) = 0\ . \tag{36}$$

Additional relations follow by tracing the change of R as the source is moved from $S_1$ to $S_2$, as shown in Figure 2.

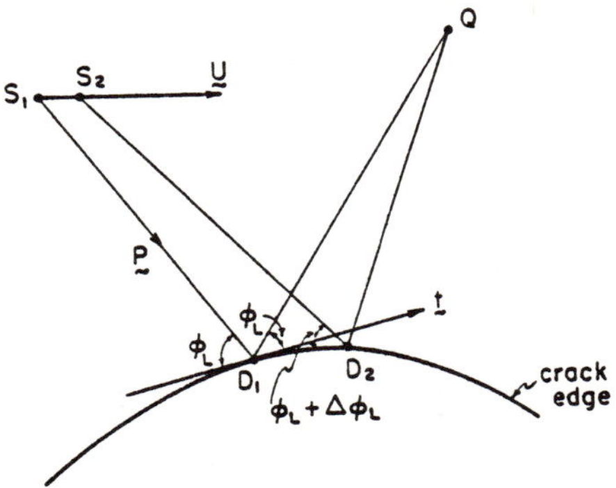


Figure 2. Change of ray path corresponding to an incremental shift of source position.

We have

$$\Delta R = \overline{S_2D_2} + \overline{D_2Q} - \overline{S_1D_1} - \overline{D_1Q} \tag{37}$$

To first order this relation can be reduced to

$$\Delta R = -\overline{S_1S_2}\ \underline{p}\cdot\underline{U} + \overline{D_1D_2}\cos(\phi_L + \Delta\phi_L) - D_1D_2\cos\phi_L, \tag{38}$$

from which it follows that

$$\underline{p}\cdot\underline{U} + \nabla_{\underline{U}}(R) = 0 . \tag{39}$$

In a similar manner we obtain

$$\underline{q}\cdot\underline{V} + \nabla_{\underline{V}}(R) = 0 . \tag{40}$$

The system of Equations 34-35 and 39-40 must be supplemented by

$$|\underline{p}| = 1 \quad \text{and} \quad |\underline{q}| = 1 \tag{41a,b}$$

The unit vector $\underline{p}$ can be solved from Equations 34, 39 and 41a. Equations 35 and 39 represent planes, while Equation 41a is a unit sphere. The intersection of each plane with the unit sphere provides a circle. The two circles have two points of intersection.

The simultaneous equations are of the general form

$$\underline{p}\cdot\underline{a} + C_1 = 0; \quad \underline{p}\cdot\underline{U} + C_2 = 0, \quad |\underline{p}| = 1 \tag{42}$$

where

$$\underline{a} = \nabla_{\underline{V}}(\underline{X}/W)/|\nabla_{\underline{V}}(\underline{X}/W)| , \quad W = R - |\underline{X}|^2 \tag{43}$$

$$C_1 = -\nabla_{\underline{V}}(R/W)/|\nabla_{\underline{V}}(\underline{X}/W)| \tag{44}$$

$$C_2 = \nabla_{\underline{U}}(R) . \tag{45}$$

It is noted that

$$|\underline{a}| = |\underline{U}| \equiv 1, \quad -1 \leq C_1, C_2 \leq 1 , \tag{46}$$

and we assume that $\underline{a}$ and $\underline{U}$ are not parallel. The solution to Equation 42 may be written as

$$\underline{p}^{\pm} = \{(C_2\underline{a} - C_1\underline{U})\wedge(\underline{a}\wedge\underline{U}) \pm [|\underline{a}\wedge\underline{U}|^2 - |C_2\underline{a} - C_1\underline{U}|^2]^{1/2} \times \times (\underline{a}\wedge\underline{U})\}/|\underline{a}\wedge\underline{U}|^2 . \tag{47}$$

In the same manner we can solve for $\underline{q}$ from Equations 36, 40 and 41b. The four solutions will have one point $\underline{x}_D$ in common, which is the point that is being sought. The position $\underline{x}_D$ may be expressed as

$$\underline{x}_D = \underline{x}_S + R_i\underline{p} \ . \tag{48}$$

In summary, then, the crack-edge mapping proceeds as follows. We start with a source point $S_1$, defined by $\underline{x}_S$, and a point of observation $Q_1$, defined by $\underline{x}_Q$, where $\underline{X}_1 = \underline{x}_Q - \underline{x}_S$. For these two points we have a total ray length R. We then define unit vectors $\underline{U}$ and $\underline{V}$ at $S_1$ and $Q_1$ respectively, and we consider a source point $S_2$ and a point of observation $Q_2$ at distances $\Delta u$ and $\Delta v$ along $\underline{U}$ and $\underline{V}$, respectively. Then, $\underline{X}_2 = \underline{X}_1 + \Delta v \ \underline{V} - \Delta u \ \underline{U}$. For $S_2$ and $Q_2$ we find the total ray length $R + \Delta R$. We can then make first order computations of the gradients appearing in Equations 43-45. Substitution of the results in Equation 47 yields the position of the flashpoint. This flashpoint corresponds to the source $S_1$ and the observer $Q_1$. The purpose of $S_2$ and $Q_2$ is to provide information for the computation of gradients. For the computation of a single flashpoint we thus require first arrival times for a pair of neighboring source positions and a pair of neighboring observer positions. A conjugate combination, for example, $S_1$ and $Q_2$ as primary positions and $S_2$ and $Q_1$ as neighboring positions can be used to check the result.

Suppose we consider one pair of source positions and N pairs of observer positions. Each pair of observer positions provides a flashpoint. Two adjacent flashpoints define the local tangent to the crack edge. Three not necessarily adjacent flashpoints determine the plane of a flat crack. Knowing the plane of the crack, we can determine the tangent at each flashpoint by the use of (1) or (10). In order that the flashpoints will not cluster too closely, and in order that a sizeable segment of the crack edge will be mapped, the pairs of observation points should be at some distance from each other.

In an alternative method for determining the flashpoint, the position of the source is kept unchanged, but a total of three observer positions are employed. In addition to the source position Q, we consider source positions $Q_1$ and $Q_2$, which are short distances from Q along the unit vectors $\underline{V}_1$ and $\underline{V}_2$, respectively, where $\underline{V}_1 \wedge \underline{V}_2 \neq \underline{0}$. According to Equation 35, the unit vector $\underline{p}$ can then be solved from

$$\underline{p}\cdot\nabla_{\underline{V}_j}(\underline{X}/W) - \nabla_{\underline{V}_j}(R/W) = 0 \quad , \quad j = 1,2 \ . \tag{49}$$

Here W is defined by Equation 33b. It follows from Equation 40 that the unit vector $\underline{q}$ can be solved from

$$\underline{q}\cdot\underline{V}_j + \nabla_{\underline{V}_j}(R) = 0 \ , \quad j = 1,2 \ . \tag{50}$$

Equations 49 and 50 each have two solutions. Of these solutions we select the $\underline{p}$ and $\underline{q}$ which give the same flashpoint.

Similarly, the flashpoint can be determined by keeping the observer fixed and moving the source in two non-parallel directions. Therefore, in terms of discrete measurements, the following combinations can be used in order to determine <u>one</u> flashpoint:

(a) One source and three non-collinear observers,

(b) Two sources and two observers, the four being non-collinear,

(c) Three non-collinear sources and one observer.

All of the above analyses apply equally well to either of the two travel distances $R_1$ and $R_2$ which are determined for each source-observer pair. Therefore each set of four measurements determines <u>two</u> distinct flashpoints. Typically, the two flashpoints lie at diametrically opposite sides of the crack, and their distance apart defines a typical diameter length.

Example

In this example we invert synthetic data computed for an elliptical crack. We nondimensionalize quantities by taking $c_L$ equal to unity.

Let us consider an ellipse in the plane z = 0, defined by

$$x^2 + 4y^2 - 1 = 0 \ , \ z = 0 \ . \tag{51}$$

Thus, the major axis is two, and the minor axis is one.

We consider three transducer positions defined by

$$1{:}(9,1,10); \ 2{:}(-3,7,10); \ 3{:}(-15,13,10) \tag{52}$$

These three points serve alternatively as source and observer positions. The notation (1,3) means that point 1 is the source and point 3 is the observer. In Figures 3 and 4 we have plotted the high frequency spectrum $V(\omega)$ of Equation 22, for the two source-observer pairs (1,3) and (1,2). In calculating $V(\omega)$, we have assumed that $U^i(\underline{x}(\psi_\alpha)) = 1$, $\alpha = 1,2$ in Equation 21. A value of 1/3 which corresponds to aluminum, is taken for Poisson's ratio. We note that Figures 3 and 4 both display the same oscillatory behavior superimposed on an

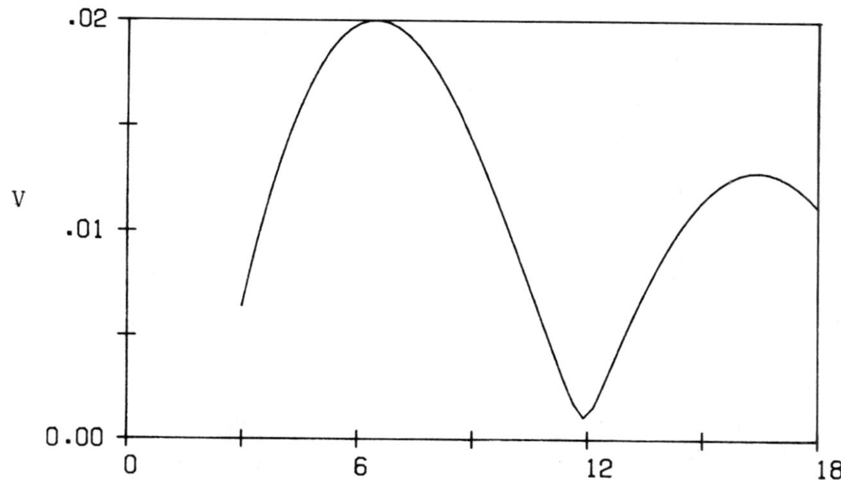


Figure 3. High frequency spectrum $V(\omega)$ of diffracted longitudinal field. Source at (9,1,10), observer at (-15,13,10).

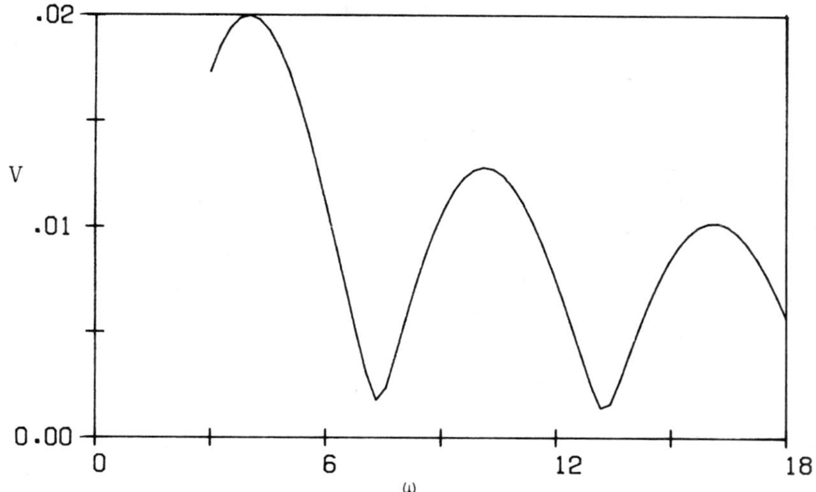


Figure 4. High frequency spectrum $V(\omega)$ of diffracted longitudinal field. Source at (9,1,10), observer at (-3,7,10).

Table 1. Comparison of exact and computed flashpoints.

| $\Delta v$ | Source Observer | Flashpoint | Computed Point | Error |
|---|---|---|---|---|
| .5 | (1,2) | ( .7570, .3267,.0) | ( .7538, .3274,-.0025) | .00411 |
| | | (-.8442,-.2680,.0) | (-.8474,-.2672,-.0023) | .00404 |
| .5 | (1,3) | (-.0413, .4996,.0) | (-.0434, .4988,-.0025) | .00334 |
| | | ( .0040,-.5000,.0) | ( .0011,-.4996,-.0034) | .00453 |
| .5 | (2,3) | (-.8264, .2816,.0) | (-.8289, .2808,-.0038) | .00461 |
| | | ( .8671,-.2490,.0) | ( .8646, .2502,-.0039) | .00475 |
| 2. | (1,2) | ( .7570, .3267,.0) | ( .7065, .3383,-.0404) | .06570 |
| | | (-.8442,-.2680,.0) | (-.8958,-.2547,-.0363) | .06453 |
| 2. | (1,3) | (-.0413, .4996,.0) | (-.0753, .4872,-.0393) | .05345 |
| | | ( .0040,-.5000,.0) | (-.0430,-.4946,-.0545) | .07218 |
| 2. | (2,3) | (-.8264, .2816,.0) | (-.8673, .2688,-.0602) | .07388 |
| | | ( .8671,-.2490,.0) | ( .8272,-.2670,-.0619) | .07582 |

inverse square root decay.

At each point a unit direction vector is defined by $(1,1,-1)/\sqrt{3}$. This corresponds to the unit vector $\underline{U}$ or $(\underline{V})$. Two values of the intertransducer distance $\Delta u$ (or $\Delta v$) have been considered. Thus, for each intertransducer distance, we have three pairs of transducers at some distance from each other. The results are presented in Table 1. The error is the distance between the actual flashpoint and the computed point. Very good agreement is achieved for an intertransducer distance of $\Delta v = .5$, but acceptable agreement is still found for $\Delta v = 2$. Whatever value of $\Delta v$ is taken, the flashpoints define a significant segment of the crack edge.

## ACKNOWLEDGMENT

The work presented here was carried out in the course of research sponsored by Ames Laboratory, for the Advanced Research Project Agency and the Air Force Materials Laboratory under Contract SC-81-005.

## REFERENCES

1. Achenbach, J. D., Viswanathan, K. and Norris, A. "An Inversion Integral for Crack-Scattering Data," *Wave Motion* 1(4):299-316 (1979).

2. Achenbach, J. D. and Gautesen, A. K. "Geometrical Theory of Diffraction for Three-D Elastodynamics," *J. Acoust. Soc. Am.* 61:413-421 (1976).

3. Gautesen, A. K., Achenbach, J. D. and McMaken, H. "Surface-wave Rays in Elastodynamic Diffraction by Cracks," *J. Acoust. Soc. Am.* 63:1824-1831 (1978).

CHAPTER 5

# SCATTERING OF WAVES BY OBSTACLES IN THE PRESENCE OF NEARBY BOUNDARIES - A SELF-CONSISTENT T-MATRIX APPROACH

*Vasundara V. Varadan and Vijay K. Varadan*
Wave Propagation Group
Department of Engineering Mechanics
The Ohio State University
Columbus, Ohio 43210

## ABSTRACT

The scattering of acoustic, electromagnetic and elastic waves by obstacles that are embedded in the presence of nearby boundaries may conveniently be studied without invoking integral representations of the field. The procedure is to use the known scattering properties of the boundary and the obstacle in terms of reflection (R-) matrices and transition (T-) matrices and account for multiple scattering processes in a systematic, self-consistent scheme. This approach would involve the use of transformation matrices that relate the basis functions used to expand the fields scattered by the boundary and the obstacle. Earlier formulations have relied on the use of integral representations to arrive at the final results. The formulation that we propose has the advantage of being applicable to acoustic, electromagnetic and elastic wave scattering by just allowing for one, two and three polarizations, respectively, in the vector decomposition of the field. For the special case in which the scalar wave equation governs the embedding medium, then the method of images may be used in a straight forward manner. A discussion of this approach is also given.

## INTRODUCTION

In many practical applications it is important to understand the effect of nearby boundaries on the scattering properties of obstacles. Even the introduction of a plane boundary, i.e. embedding the obstacle in a half space rather than in an infinite medium causes a significant change in the scattered field and at the same time presents a mathematical challenge except for the case of scalar waves. The most straight forward approach for problems involving obstacles in half spaces is to use a problem oriented Green's function namely that of the half space. Only for the scalar case the half space Green's function is known analytically. For the vector electromagnetic and elastic case only an integral representation of the half space Green's function is known. Further for the case of scalar waves, the use of the half space Green's function leads in a straight forward manner to the method of images as discussed in section 4.

Obstacles in half spaces are of two types, one in which the obstacle is a closed one and is completely below the plane Γ bounding the half space (Figure 1) or one could consider the problem of a surface indentation of Γ otherwise known as a surface breaking crack in the mechanics literature (see Figure 2). In electromagnetics the problem could be

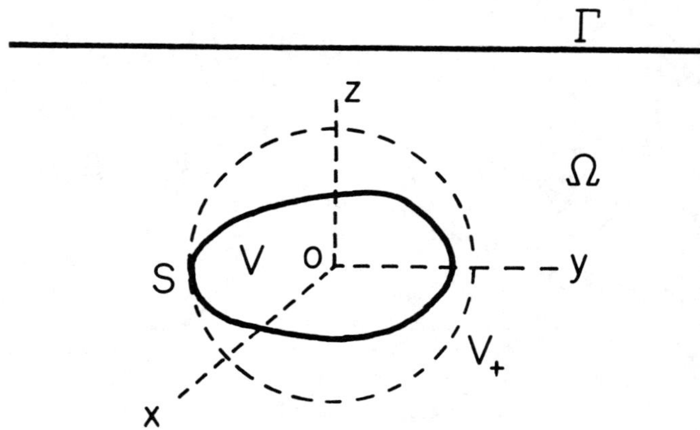


Figure 1. A three-dimensional scatterer in an elastic half space.

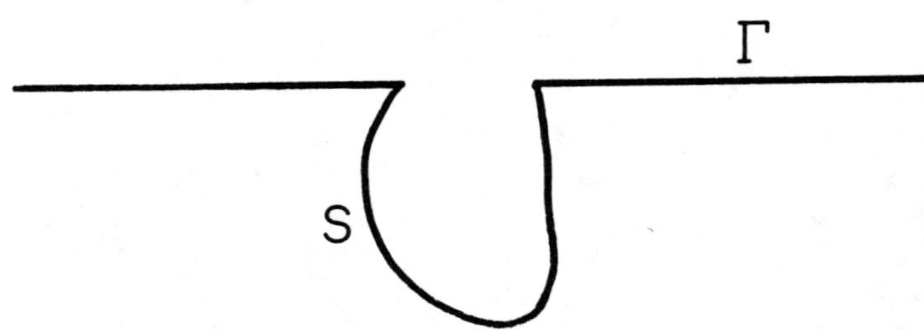


Figure 2. A three-dimensional surface indentation.

that of the scattering characteristics of an antenna mounted on a perfectly conducting plane. If one uses the half space Green's function, both types of problems can be treated automatically but some of the other approaches are not applicable to the problem of surface indentations.

The diffraction of scalar waves in 2-D by a buried cylinder as well as a surface slit has been studied by Datta et al. [1,2]. Stone, Ghosh and Mal [3] have studied the diffraction of scalar waves by surface slits using integral equations. Mendelsohn, Achenbach and Keer [4] have studied the two-dimensional vector elastic wave scattering by surface slits. Kristenssön and Ström [5], and Boström and Kristenssön [7] have given a formulation based on the use of the T-matrix approach (null field method) and integral representations to study acoustic, electromagnetic and elastic wave scattering by buried obstacles. In References 5-7, numerical results have been obtained for the scattering of waves by non-spherical obstacles embedded in half spaces. Recently Bostrom, Wall, Varadan and Varadan [8] have studied the problem of elastic wave scattering by a 3-D surface breaking flaw using a problem oriented Green's function and the T-matrix approach.

In this paper we present a self-consistent multiple scattering approach to deal with the boundaries that may be present near an obstacle. We have pursued a similar approach when studying the multiple scattering of waves by a finite number of scatterers [9] as well as a statistical ensemble of scatterers [10]. The basic idea is rather simple. Let the total field at a point be $u^{tot} = u_E^i + u_S^i$, where $u_E^i$ is the field exciting or incident on the i-th scatterer (the scatterer could be a closed obstacle or a boundary such as an infinite plane) and $u_S^i$ is the field scattered by it. The exciting field $u_E^i$ in turn contains not only the primary incident wave but also the waves scattered by all the other scatterers. Then one can expand $u_E^i$ and $u_S^i$ in a suitable set of basis functions with unknown expansion coefficients $\alpha$ and $\beta$ respectively. For closed obstacles $\beta$ is related to $\alpha$ via the T-matrix [11,12,13,14]. For infinite boundaries, plane or otherwise $\alpha$ and $\beta$ are related via the reflection matrix. Since the reflection matrix of a general boundary involves the inversion of a two dimensional integral transform, it is difficult to obtain explicit results except in the case of a plane boundary. After defining expansions for $u_E^i$ and $u_S^i$ and relating $\alpha$ and $\beta$ via the R- or T-matrices, one simply ends up with a system of coupled matrix equations which may be solved in a straight forward manner.

This method is rather attractive due to its basic simplicity. The only requirement is that the T-matrices and R-matrices for the given scatterer and boundary must be evaluated using the appropriate boundary conditions. It is in the evaluation of R and T that explicit integral representations of the fields must be involved. Otherwise, the multiple scattering formalism given in Section 2 is applicable to acoustic, electromagnetic and elastic fields if the proper set of basis functions appropriate to the particular field are used. Unlike the approach used in References 5-7, extension to the case of an obstacle between two parallel boundaries is quite straight forward. One of the drawbacks of this method is that one cannot include the case of a surface indentation since one cannot define the T-matrix of the indentation, which is an open surface in the usual manner. This problem was overcome in Reference 8 by the use of the half space Green's function. But the final expression for the total scattered field is the same in References 7 and 8 as well as the present formulation if the T-matrix of an indentation is only introduced in a formal sense. This T-matrix is no longer symmetric as is the T-matrix of a closed obstacle. If one is willing to accept this formal but unorthodox definition of T for an indentation then Section 2 of this paper is applicable to surface indentations as well.

Finally, in Section 4 we discuss the more simple problem of scalar wave scattering by an obstacle in a half space by the method of images. In appendix A and B, the plane and spherical vector wave functions are defined, the transformations between them introduced and the translation matrices for scalar spherical functions needed for the method of images are given.

## A SELF-CONSISTENT MULTIPLE SCATTERING FORMALISM

Consider a semi-infinite homogeneous, isotropic medium bounded by the plane $\Gamma$ in which an obstacle bounded by the surface S is embedded, see Figure 1. A plane harmonic wave of frequency $\omega$ is incident on this geometry. The field at any point in the medium is the sum of the incident wave field plus the fields due to waves scattered by the plane $\Gamma$ and the obstacle, which are denoted by $\vec{u}^0$, $\vec{u}_S^\Gamma$ and $\vec{u}_S^S$, respectively. Thus,

$$\vec{u}(\vec{r}) = \vec{u}^0(\vec{r}) + \vec{u}_S^\Gamma(\vec{r}) + \vec{u}_S^S(\vec{r}) \; . \tag{1}$$

The field that excites or is incident on the obstacle is the incident wave field plus the waves reflected from the plane which can be written as follows:

$$\vec{u}_E^S(\vec{r}) = \vec{u}^0(\vec{r}) + \vec{u}_S^\Gamma(\vec{r}) \; ; \; \vec{r} \in V_+ \tag{2}$$

where $\vec{r}$ is a point in the region $V_+$ enclosed between S and the sphere that circumscribes S. Similarly, for the field that excites the plane $\Gamma$, we have

$$\vec{u}_E^\Gamma(\vec{r}) = \vec{u}^0(\vec{r}) + \vec{u}_S^S(\vec{r}) \; ; \; \vec{r} \in \Gamma \; . \tag{3}$$

The total scattered field is then given by

$$\vec{u}_S^{tot}(\vec{r}) = \vec{u}_S^\Gamma(\vec{r}) + \vec{u}_S^S(\vec{r}) \; . \tag{4}$$

Our aim is to relate the total scattered field to the incident wave field by expanding all fields in a suitable set of basis functions. A convenient choice is to use expansions in plane waves for fields incident on and scattered by the plane and to use spherical wave functions to expand fields incident on and scattered by the obstacle. Since the plane wave fields and spherical wave fields occur in the same equations, it becomes necessary to use the transformation properties between plane waves and spherical waves. The transformation properties of vector and scalar, plane and spherical waves have been reviewed by Boström, Kristensson and Ström [15]. The results are summarized in Appendix A.

The fields are expanded as follows using 0 in V as the origin of a coordinate system (see Figure 1). The time dependence of all fields which is of the form exp(-iωt) is not written explicitly. The incident field can be represented as an expansion in either the spherical wave basis or the plane wave basis

$$\vec{u}^0(\vec{r}) = \sum_n a_n \; \mathrm{Re} \; \vec{\psi}_n(\vec{r}) \tag{5a}$$

$$= \sum_j \int_{C_+} \sin\alpha \; d\alpha \int_0^{2\pi} d\beta \; a_j(\hat{\gamma}) \; \vec{\phi}_j(\hat{\gamma},\vec{r}) \tag{5b}$$

where $\alpha,\beta$ are spherical polar angles and $\hat{\gamma} = (\sin\alpha\cos\beta, \sin\alpha\sin\beta, \cos\alpha)$, $\hat{\alpha}$ and $\hat{\beta}$ are unit vectors of a spherical polar coordinate system. The subscript n in Equation 5a is a multiindex representing the different possible polarizations as well as the orbital and azimuthal modes of spherical waves. The summation on j in Equation 5b is over the different polarizations of the incident field. The coefficients of expansion $a_n$ and $a_j$ are assumed to be known.

The fields associated with the plane Γ are also expanded in a plane wave basis as follows

$$\vec{u}_E^\Gamma(\vec{r}) = \sum_j \int_{C_+} \sin\alpha \; d\alpha \int_0^{2\pi} d\beta \; \alpha_j(\hat{\gamma}) \; \vec{\phi}_j(\hat{\gamma},\vec{r}) \tag{6}$$

$$\vec{u}_S^\Gamma(\vec{r}) = \sum_j \int_{C_-} \sin\alpha \; d\alpha \int_0^{2\pi} d\beta \; \beta_j(\hat{\gamma}) \; \vec{\phi}_j(\hat{\gamma},\vec{r}) \tag{7}$$

where $C_+$ and $C_-$ are the contours as defined in Figure 3, see also Appendix A, and the coefficients $\alpha_j$ and $\beta_j$ are unknown.

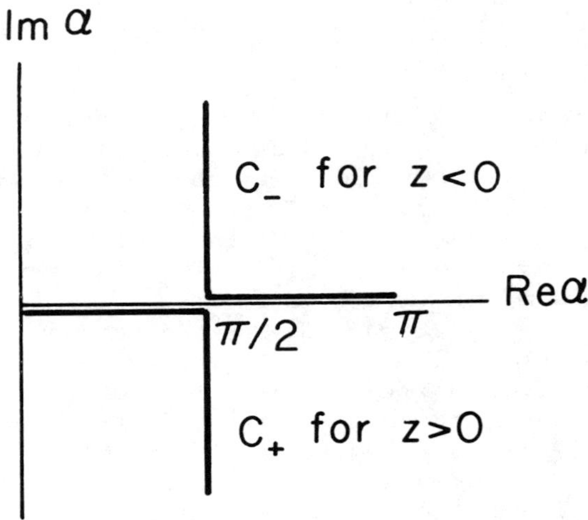


Figure 3. Integration contours $C_+$ and $C_-$.

Similarly, the fields associated with the scatterer are expanded in the spherical wave basis as

$$\vec{u}_E^S(\vec{r}) = \sum_n \alpha_n \operatorname{Re} \vec{\psi}_n(\vec{r}) \quad ; \quad a \leq |\vec{r}| \leq 2a \tag{8}$$

$$\vec{u}_S^S(\vec{r}) = \sum_n \beta_n \vec{\psi}_n(\vec{r}) \quad ; \quad |\vec{r}| > a \tag{9}$$

where $\alpha_n$ and $\beta_n$ are unknown coefficients.

Expansions given in Equations 5-9 are substituted in Equations 2 and 3 which yield

$$\sum_j \int_{C_+} \sin\alpha \, d\alpha \int_0^{2\pi} d\beta \; \alpha_j(\vec{r}) \; \vec{\phi}_j(\hat{\gamma},\vec{r})$$

$$= \sum_j \int_{C_+} \sin\alpha \, d\alpha \int_0^{2\pi} d\beta \; a_j(\hat{\gamma}) \; \vec{\phi}_j(\hat{\gamma},\vec{r}) + \sum_n \beta_n \vec{\psi}_n(\vec{r}) \tag{10}$$

$$\sum_n \alpha_n \operatorname{Re} \vec{\psi}_n(\vec{r}) = \sum_n a_n \operatorname{Re} \vec{\psi}_n(\vec{r}) +$$

$$+\sum_j \int_{C_-} \sin\alpha\, d\alpha \int_0^{2\pi} d\beta\; \beta_j(\hat{\gamma})\; \hat{\phi}_j(\hat{\gamma},\vec{r}) \;. \tag{11}$$

In Equation 10, the spherical functions $\vec{\psi}_n(\vec{r})$ are transformed to the plane wave basis and in Equation 11, the plane waves are transformed into spherical waves using Equations A-1 through A-12. Using the orthogonality of the plane waves as well as the spherical waves on the surface of a sphere of constant radius = $|\vec{r}|$, we obtain

$$\alpha_j(\vec{r}) = a_j(\hat{\gamma}) + 2i \sum_n \beta_n\, B_{nj}(\hat{\gamma}) \tag{12}$$

and

$$\alpha_n = a_n + (-i) \sum_j \int_{C_-} \sin\alpha\, d\alpha \int_0^{2\pi} d\beta\; \beta_j(\hat{\gamma})\; B_{nj}^{\dagger}(\hat{\gamma}) \;. \tag{13}$$

At this point in the formulation, we introduce the concept of a T-matrix for the obstacle and a reflection matrix R for the plane Γ. The T-matrix as defined using the null field equation relates the expansion coefficients of the field incident on an obstacle to the expansion coefficients of the field scattered by it [11,12,13]. Similarly, the reflection matrix R relates the coefficients of expansion of this field exciting the plane surface Γ to the coefficients of the field scattered by it [5,6,7]. Thus, we have

$$\beta_n = \sum_{n'} T_{nn'}\, \alpha_{n'} \tag{14}$$

and

$$\beta_j(\hat{\gamma}) = \sum_{j'} \int_{C_+} \sin\alpha'\, d\alpha' \int_0^{2\pi} d\beta'\; R_{jj}(\hat{\gamma},\hat{\gamma}')\; \alpha_{j'}(\hat{\gamma}')$$

$$= \sum_{j'} \int_{C_+} d\hat{\gamma} R_{jj'}(\hat{\gamma},\hat{\gamma}')\; \alpha_{j'}(\hat{\gamma}') \tag{15}$$

In Equation 15 and what follows, we shall use the abbreviation

$$\int_{C_\pm} d\hat{\gamma} = \int_{C_\pm} \sin\alpha\, d\alpha \int_0^{2\pi} d\beta$$

Using Equations 14 and 15 in 12 and 13, we obtain

$$\alpha_j(\hat{\gamma}) = a_j(\hat{\gamma}) + 2i \sum_j \sum_{j'} B_{nj}(\hat{\gamma})\, T_{nn'}\, \alpha_{n'} \tag{16}$$

$$\alpha_n = a_n - i \sum_j \sum_{j'} \int_{C_-} d\hat{\gamma} \int_{C_+} d\hat{\gamma}'\, B^{\dagger}_{nj}(\hat{\gamma})\, R_{jj'}(\hat{\gamma},\hat{\gamma}')\, \alpha_{j'}(\hat{\gamma}') \tag{17}$$

Equations 16 and 17 yield

$$\alpha_n = a_n - i \sum_j \sum_{j'} \int_{C_-} d\hat{\gamma} \int_{C_+} d\hat{\gamma}'\, B^{\dagger}_{nj}(\hat{\gamma})\, R_{jj'}(\hat{\gamma},\hat{\gamma}')\, a_{j'}(\hat{\gamma}')$$

$$+ 2 \sum_{n'} \sum_{n''} \sum_j \sum_{j'} \int_{C_-} d\hat{\gamma} \int_{C_+} d\hat{\gamma}'\, B^{\dagger}_{nj}(\hat{\gamma})\, R_{jj'}(\hat{\gamma},\hat{\gamma}')\, B_{n'j'}(\hat{\gamma}')\, T_{n'n''}\alpha_{n''} \tag{18}$$

The reflection matrix $R_{jj'}(\hat{\gamma},\hat{\gamma}')$ which is calculated in a plane wave basis may be expressed in a spherical wave basis by the following transformation:

$$R_{nn'} = 2 \sum_j \sum_{j'} \int_{C_-} d\hat{\gamma} \int_{C_+} d\hat{\gamma}'\, B^{\dagger}_{nj}(\hat{\gamma})\, R_{jj'}(\hat{\gamma},\hat{\gamma}')\, B_{n'j'}(\hat{\gamma}'). \tag{19}$$

Using the above definition in the last term of Equation 18, we have

$$\alpha_n = a_n - i \sum_j \sum_{j'} \int_{C_-} d\hat{\gamma} \int_{C_+} d\hat{\gamma}'\, B^{\dagger}_{nj}(\hat{\gamma})\, R_{jj'}(\hat{\gamma},\hat{\gamma}')\, a_j(\hat{\gamma}')$$

$$+ \sum_{n'} \sum_{n''} R_{nn'}\, T_{n'n''}\, \alpha_{n''} \tag{20}$$

The first two terms of Equation 20 represent the direct plane wave and the spectrum of plane waves reflected from the plane $\Gamma$ that reach the obstacle. For brevity, we introduce the notation $d_n$ (in accordance with Reference 7) to express the expansion coefficients of the plane waves incident on the obstacle. In vector matrix notation, Equation 20 may be solved to yield

$$\alpha = [1-RT]^{-1}\, d \tag{21}$$

where

$$d_n = a_n - i \sum_j \sum_{j'} \int_{C_-} d\hat{\gamma} \int_{C_+} d\hat{\gamma}'\, B^{\dagger}_{jn}(\hat{\gamma})\, R_{jj}(\hat{\gamma},\hat{\gamma}')\, a_{j'}(\hat{\gamma}') \tag{22}$$

We have now formally completed the solution of the scattering of waves by an obstacle embedded in a half space. Using Equations 14-16 and 21-22 in Equations 4, 7 and 9, we obtain

$$u_S^{tot}(\vec{r}) = \sum_j \sum_{j'} \int_{C_-} d\hat{\gamma}\, \vec{\phi}_j(\hat{\gamma},\vec{r}) \int_{C_+} d\hat{\gamma}'\, R_{jj'}(\hat{\gamma},\hat{\gamma}')\, a_{j'}(\hat{\gamma}')$$

$$+ 2i \sum_n \sum_{n'} \sum_\gamma \sum_{\gamma'} \int_{C_-} d\hat{\gamma}\, \vec{\phi}_j(\hat{\gamma},\vec{r}) \int_{C_+} d\hat{\gamma}' R_{jj'}(\hat{\gamma},\hat{\gamma}') B_{nj'}(\hat{\gamma}') T_{nn'} \alpha_{n'}$$

$$+ \sum_n \sum_{n'} \vec{\psi}_n(\vec{r})\, T_{nn'} \alpha_{n'} \quad . \qquad (23)$$

The first term in the expression for the total scattered field at an observation point $\vec{r}$ is simply the waves reflected by the plane $\Gamma$. The next two terms account for the multiple reflections between the plane and the obstacle. The second term in Equation 23 is the multiply scattered wave that is radiated by the plane and the third term is the multiply scattered wave emanating from the obstacle. It is the term $[1-RT]^{-1}$ in Equation 21 that when expanded formally displays the various orders of multiple scattering between the plane and the obstacle. These comments may be clearly seen with the aid of Figure 4.

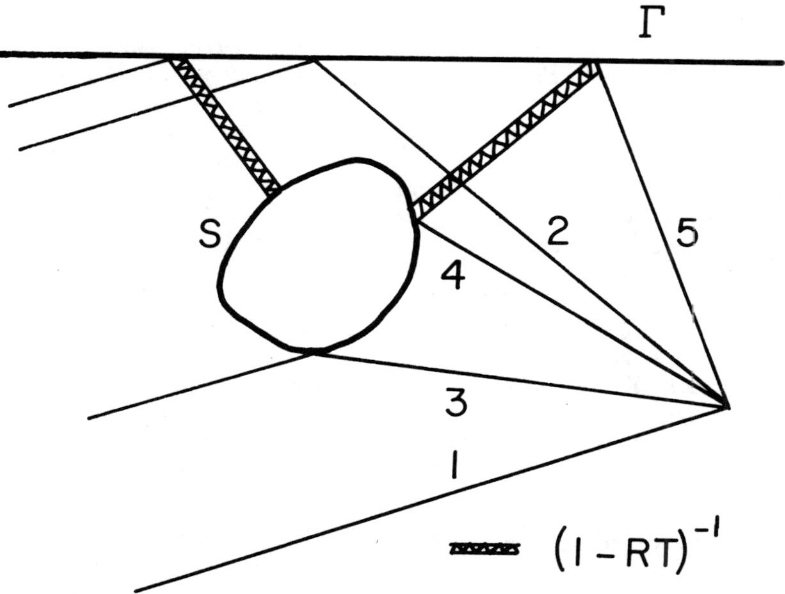


Figure 4. Different ray paths to the field at P.

The final expression for the total scattered field, Equation 23, is in complete agreement with the work of Kristensson and Ström [6] and Boström and Kristensson [7] for scalar, vector electromagnetic and elastic wave scattering by obstacle embedded in a half space. Their work is based on the use of integral representations of field and hence a separate formalism is required for the three types of fields. The advantage of the present approach is that once the T-matrix of the obstacle is obtained from the integral representation of the appropriate field, then Equation 23 can be used for all three fields by keeping the required polarization index j and $\tau$.

Recently, Boström, Wall, Varadan and Varadan [8] have considered the problem of elastic wave scattering from a surface breaking crack by using the Green's function for an elastic half space. The expression for the total scattered field in this case is also the same as Equation 23. However, within the present formulation it is difficult to include the surface breaking crack as a special case since it is not possible to define the T-matrix of an open obstacle in the usual manner. In Reference 8, such an expression occurs naturally although the T-matrix thus defined does not possess any symmetry or unitary properties as does the T-matrix of a closed obstacle. The geometrical restrictions in the present formulation for buried obstacles is that the sphere (3-D) or circle (2-D) circumscribing the obstacle should not extend into the region above $\Gamma$.

Details of obtaining the reflection matrix R and the expansion coefficients $d_n$ of Equation 22 are given in analytic form in References 5-8. The details of evaluating the T-matrix of the obstacle are given by Waterman [11-13] and Varadan and Pao [14] for acoustic, electromagnetic and elastic waves. We observe that the obstacle may be rather general since the boundary conditions imposed at S are contained in T. It may be homogeneous or a layered obstacle and even be replaced by a configuration of obstacles as long as the geometrical restrictions of the previous paragraph are not violated.

## OBSTACLES BETWEEN TWO PARALLEL BOUNDARIES

It is now fairly straight forward to consider the scattering of waves by obstacles between two parallel boundaries as shown in Figure 5. Let $\Gamma^+$ be the upper plane and $\Gamma^-$ the lower plane. The total scattered field is now given by

$$\vec{u}_S^{tot}(\vec{r}) = u_S^{\Gamma+}(\vec{r}) + \vec{u}_S^{\Gamma-}(\vec{r}) + \vec{u}_S^{S}(\vec{r}) \qquad (24)$$

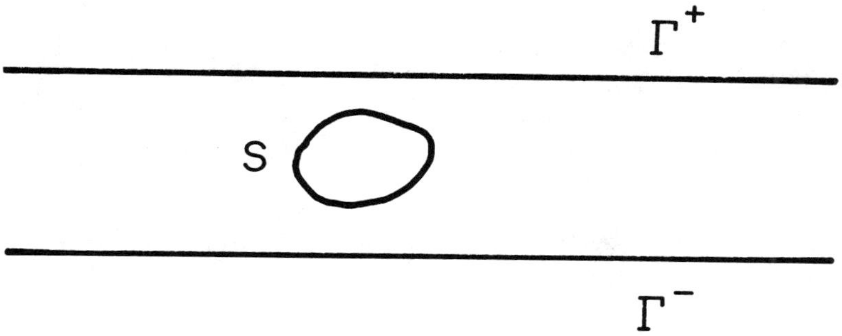

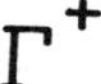

$\Gamma^-$

**Figure 5.** Obstacle in a plate.

and the exciting field on $\Gamma^+, \Gamma^-$ and S are given by

$$\vec{u}_E^{\Gamma+} = \vec{u}^0 + \vec{u}_S^{\Gamma-} + \vec{u}_S^S \tag{25}$$

$$\vec{u}_E^{\Gamma-} = \vec{u}^0 + \vec{u}_S^{\Gamma+} + \vec{u}_S^S \tag{26}$$

and

$$\vec{u}_E^S = \vec{u}^0 + \vec{u}_S^{\Gamma+} + \vec{u}_S^{\Gamma-} \tag{27}$$

We may use the same expansions for $\vec{u}_E^{\Gamma+}$, $\vec{u}_S^{\Gamma+}$, $\vec{u}_E^S$ and $\vec{u}_S^S$. For plane $\Gamma^-$, the following expansions are appropriate

$$\vec{u}_E^{\Gamma-} = \sum_j \int_{C_-} d\hat{\gamma}\ \alpha_j^-(\hat{\gamma})\ \vec{\phi}_j(\hat{\gamma}, \vec{r}) \tag{28}$$

and

$$\vec{u}_S^{\Gamma-} = \sum_j \int_{C_+} d\hat{\gamma}\ \beta_j^-(\hat{\gamma})\ \vec{\phi}_j(\hat{\gamma}, \vec{r}) \tag{29}$$

When all the expansions are substituted into Equations 25-27, the reflection matrix $R^-$ for $\Gamma^-$ is introduced, then we get a set of 3 coupled matrix equations for $\alpha_j^+$, $\alpha_j^-$ and $\alpha_n$ which we can proceed to solve.

## METHOD OF IMAGES

In those problems for which the scalar Helmholtz equation governs wave propagation in the half space, one can successfully employ the method of images. Image theory results directly from the use of the half space Green's function in the integral representations of the field. This approach is thus applicable to sound soft, sound hard and elastic obstacles in fluid half spaces, SH-wave scattering by infinitely long obstacles in elastic half spaces and electromagnetic wave scattering by infinitely long scatterers in the presence of a perfectly conducting plane. Thus, one can find a variety of practical problems that are amenable to a solution by the method of images.

The integral representation for the field in a half space in which an obstacle is embedded is given by

$$\vec{u}^0(\vec{r}) + \int_\Gamma (u'\hat{n}'.\nabla' g(\vec{r},\vec{r}\,') - g(\vec{r},\vec{r}\,')\hat{n}.\nabla'_+ u')\, d\Gamma + \int_S (u'\hat{n}'.\nabla' g(\vec{r},\vec{r}\,') - g(\vec{r},\vec{r}\,')\,\hat{n}.\nabla'_+ u')\, dS = \begin{cases} u(\vec{r}) \; ; \; \vec{r}\ \varepsilon\ \Omega \\ 0 \; ; \; \text{otherwise} \end{cases} \tag{30}$$

In the above equation $g(\vec{r},\vec{r}\,')$ is the free space Green's function. The unit outward normal to the surfaces $\Gamma$ and S is denoted by $\hat{n}$ and primes denote that the field is a function of $\vec{r}\,'$, a point on the surface. The incident field $u^0(\vec{r})$ is any harmonic field produced by sources in $\Omega$.

In addition to boundary conditions that may be imposed on S, Neumann or Dirichlet boundary conditions may be prescribed on the plane $\Gamma$ and finally the Green's function and the scattered field must satisfy radiation conditions at infinity. For the scalar Helmholtz equation, the Green's function satisfying proper radiation conditions at infinity is given by

$$g(\vec{r},\vec{r}\,') = \begin{cases} i\pi\, H_0^{(1)}(k|\vec{r}-\vec{r}\,'|) & \text{in 2-D} \\ \exp(ik|\vec{r}-\vec{r}\,'|)/|\vec{r}-\vec{r}\,'| & \text{in 3-D} \end{cases} \tag{31}$$

where $H_0^{(1)}$ is the Hankel function of the first kind of zero order. We can easily construct a Green's function that satisfies in addition the required boundary conditions on $\Gamma$ in a very simple manner. Thus, if Neumann boundary conditions are prescribed on $\Gamma$, we require then $\hat{n}\,.\,\nabla_+ u = 0$ on $\Gamma$ and we see that

$$g^N(\vec{r},\vec{r}\,') = g(\vec{r},\vec{r}\,') + g(\vec{r},\vec{r}\,'_I) \tag{32}$$

with

$$\vec{r}\,'_I = \vec{r}\,'(x',y',-z') \; ; \; \vec{r}\,' = \vec{r}\,'(x',y',z')$$

satisfies the boundary condition

$$\hat{n}.\nabla g^N(\vec{r},\vec{r}\,') = 0 \; ; \; \vec{r} \text{ on } \Gamma \; . \tag{33}$$

Similarly, if Dirichlet boundary conditions are prescribed on $\Gamma$, then $u_+ = 0$ on $\Gamma$ and we observe that

$$g^D(\vec{r},\vec{r}') = g(\vec{r},\vec{r}') - g(\vec{r},\vec{r}'_I) \tag{34}$$

is such that

$$g^D(\vec{r},\vec{r}') = 0 \ , \ \vec{r} \text{ on } \Gamma \ . \tag{35}$$

Using these new forms for the Green's function in the integral representation of the field, we have

$$u^0(\vec{r}) + \int_S (u_+ \ \hat{n}.\nabla g^{N,D} - g^{N,D} \ \hat{n}.\nabla_+ u)\,dS = \begin{cases} u(\vec{r}) \ ; \ \vec{r} \in \Omega \\ 0 \ ; \ \text{otherwise} \end{cases} \tag{36}$$

In Equation 36, we use $g^N$ or $g^D$ depending on whether Neumann or Dirichlet boundary conditions are imposed on $\Gamma$.

We note that the point $\vec{r}'_I$ may be considered as the image of the source point $\vec{r}'$. Thus, the half plane $\Gamma$ acts like a mirror and the single surface integral in Equation 36 containing the half space Green's functions $g^N$ or $g^D$ may be replaced by two surface integrals, one over the surface S and the other over the surface $S_I$ of the image of the obstacle and replace $g^N$, $g^D$ by the free space Green's function. But for the total field in $\Omega$ to correctly satisfy the proper boundary condition on $\Gamma$, the source in $\Omega$ must be augmented by image sources in $\Omega_I$. Thus, the equivalent scattering problem using the method of images is given by Figure 6, where the + sign is used for Neumann boundary conditions on $\Gamma$ and the - sign for Dirichlet boundary conditions on $\Gamma$.

The two scatterer configurations in Figure 6 may be solved by using the T-matrix approach for configurations containing a finite number of scatterers as described by the authors elsewhere [9]. If 'd' is the depth of the obstacle below the plane $\Gamma$, then the image is at a height 'd' above the plane. The T-matrix of an obstacle in a half space is then given by

$$T^H = R(\vec{d})\ T\ [1-\sigma(2\vec{d})\ T\sigma(-2\vec{d})T]^{-1}\ [1+\sigma(2\vec{d})\ TR(-2\vec{d})]\ R(-\vec{d})$$
$$+ R(-\vec{d})\ T\ [1-\sigma(-2\vec{d})\ T\sigma(2\vec{d})\ T]^{-1}\ [1+\sigma(-2\vec{d})\ TR(2\vec{d})]\ R(\vec{d}) \tag{37}$$

where T is the usual T-matrix of the obstacle if it were embedded in an infinite medium, $\sigma$ and R are translation matrices for the scalar spherical wave functions, the expressions of which are given in Appendix B. The inverses in Equation 37 when expanded formally leads to the multiple scattering of waves between the obstacle and its image.

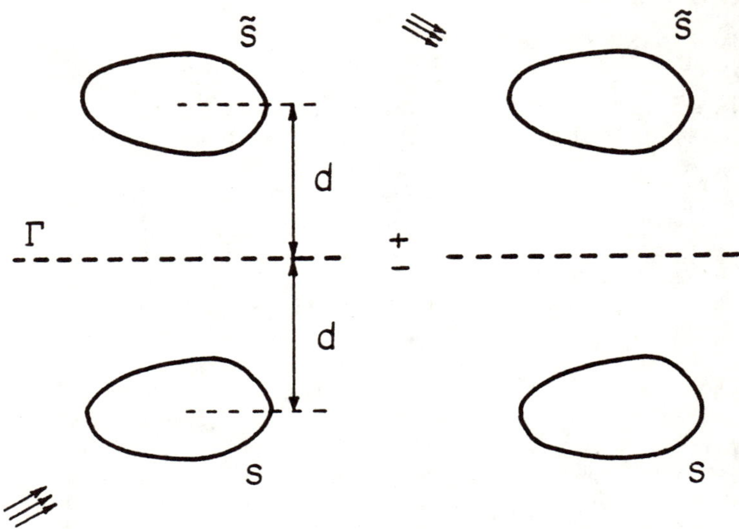


Figure 6. Equivalent scattering configuration for Neumann(+) and Dirichlet(-) boundary conditions on Γ.

It is interesting to compare the expression obtained for the T-matrix of an obstacle in a half space, keeping in mind that one must also irradiate the configuration by the image of the original sources with the expression for the total scattered field in the previous section, namely Equation 23. It becomes convenient to write an explicit equation for the scattered field in terms of $T^H$. In this case

$$u_S^{tot} = \sum_n \sum_{n'} \psi_n(\vec{r}) \; [T^H_{nn'} \, a_{n'} + T^H_{nn'} \, \overline{a}_{n'}] \tag{38}$$

where $a_n'$ are the expansion coefficients of the incident wave and $\overline{a}_n'$ that of its image.

We have remarked that in Equation 23 it is the term $[1-RT]^{-1}$ where R is the reflection matrix of the plane Γ and T is the matrix of the obstacle that accounts for multiple scattering. In the method of images, the T-matrix of the image and the image of the source take the place of the reflection matrix. With the use of image theory one complication is the appearance of the translation matrices; this is because separate coordinate systems must be used to describe fields with respect to the obstacle and its image. Whereas in the previous formalism, the same coordinate system sufficed for the plane and the obstacle.

## NUMERICAL RESULTS

Sample numerical results are provided for the case of an oblate spheroidal cavity in an elastic aluminum half space with its axis of revolution normal to the surface $\Gamma$. Traction free boundary conditions are prescribed on $\Gamma$ and S. Analytic expressions for the reflection matrix have been obtained by Boström and Kristenssön [7] and they are not repeated here. Plots of the surface displacement field on $\Gamma$ are plotted for an oblate spheroid of aspect ratio a/b = 5.0 where 'a' and 'b' are the semi-major and semi-minor axes, respectively. The center of the spheroid is at a depth $z_o/a = 2.0$ below $\Gamma$ on the z-axis. Rayleigh wave, P-wave and SV-wave incidence are considered at several different frequencies for different scattering geometries. In many applications it may be possible to measure only surface displacements as a function of frequency of the incident wave at several locations of $\Gamma$. At the present Equation 23 for the scattered field computations is not of much practical value since the numerical convergence of some of the expressions in $u_S^{tot}$ is rather poor. The surface displacement on $\Gamma$ can be computed since the coefficients of expansion $\alpha_j(\hat{\gamma})$ of $u_E^{\Gamma}$ are known. These may be used in the extinction theorem for points above $\Gamma$ to obtain expressions for the field on $\Gamma$.

The geometry used in computations is displayed in Figure 7. The field point on $\Gamma$ is denoted by its distance $\rho$ from the z-axis along the x-axis. In Figure 8, the absolute value of the x- and z-components of the surface displacement field are plotted as a function of $k_p x_o$ for a Rayleigh wave incident along the x-direction. P- and SV- waves incident along the z-axis are considered in Figure 9 and the surface displacement on $\Gamma$ directly above the center of the spheroid is plotted as a function of frequency. In this case due to symmetry, at this point $\rho/z_o = 0$, the x- component of the displacement on $\Gamma$ for P-wave incidence is zero and the z- component of the displacement on $\Gamma$ for S- wave incidence is zero. In Figure 10, the x- and z- components of the displacement field on $\Gamma$ are plotted for a point directly above the center of the spheroid ($\rho/z_o = 0$) for P- waves incident in the x-z plane at $30^o$ to the positive z-axis and for S-waves incident at $29.4^o$ with respect to the positive z-axis. This angle is just slightly smaller than the Rayleigh critical angle for the material considered (Aluminum). We observe that the z- component of the displacement for incident S-waves is much smaller throughout the whole range of frequencies.

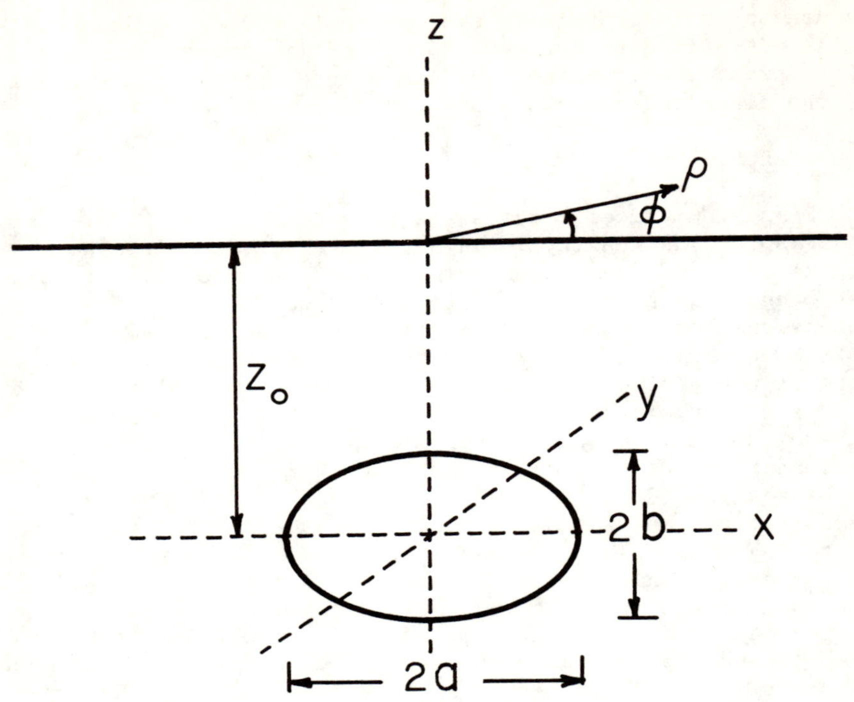


Figure 7. Oblate spheroid in an elastic half space.

It is worth remarking at this stage that all the numerical results shown here are for the surface displacement field on Γ. In order to compute the scattered field at a point in Ω far from S and Γ, one would have to resort to an asymptotic method to evaluate some of the integrals in Equation 23 analytically to avoid convergence problems. This would be of particular value in NDE applications involving the detection of flaws close to the inner wall of a thick walled vessel by transmitting and receiving waves from the outer surface.

## ACKNOWLEDGMENTS

This research was supported by the Center for Advanced NDE operated by the Ames Laboratory under Contract No. SC-81-012 and the Welding Research Center at the Ohio State University under a joint NSF-Industry Grant. The authors wish to thank Dr. Anders Boström for many valuable discussions. The assistance provided by Dr. T. Pillai and Mr. J. Tsao with computations is acknowledged.

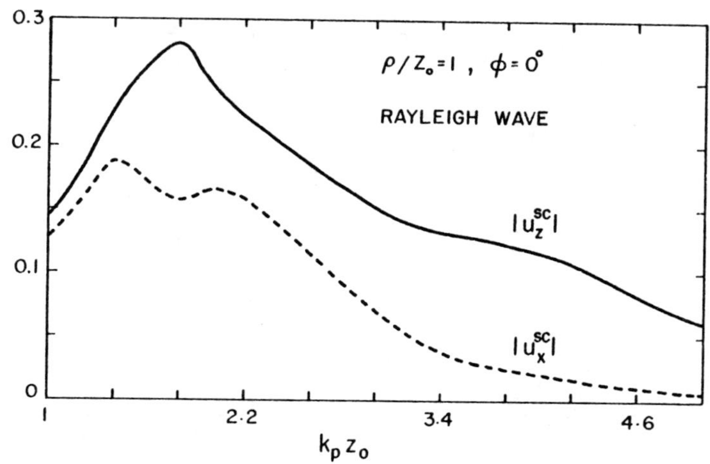


Figure 8. Surface displacement on an elastic half space containing an oblate spheroid for Rayleigh wave incidence.

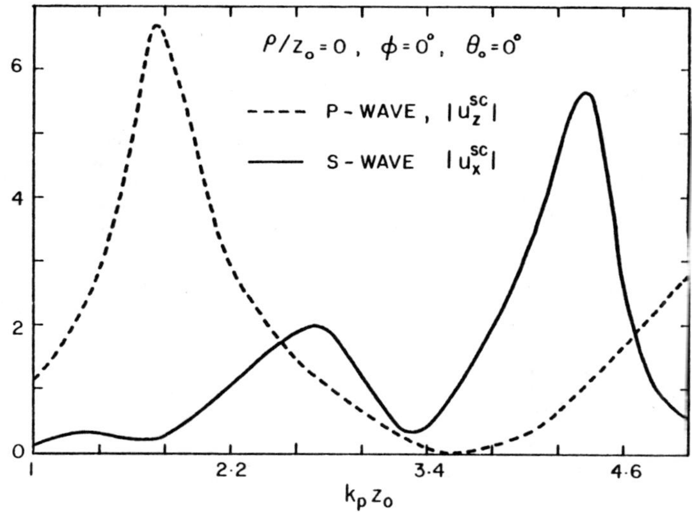


Figure 9. Surface displacement on an elastic half space containing an oblate spheroid for P- and S- wave incidence at $\theta_o = 0^o$.

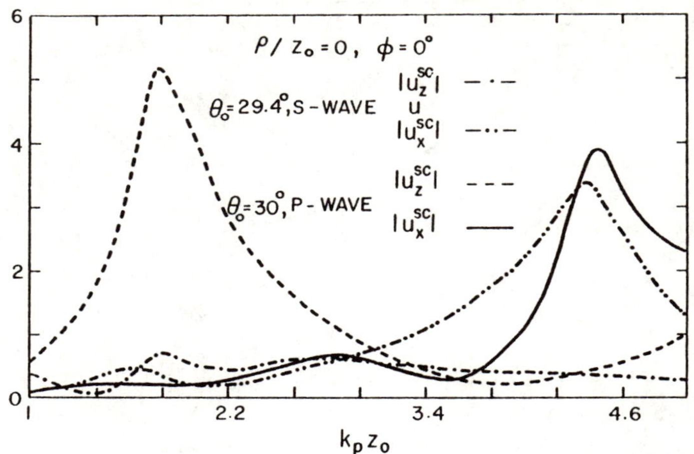


Figure 10. Surface displacement on an elastic half space containing an oblate spheroid for P-wave at $\theta_o = 30^o$ and S-wave at the critical angle, $\theta_o = 29.4^o$, of incidence.

APPENDIX A

Vector Spherical and Vector Plane Waves

In this section, we define the vector spherical and plane wave functions that were used in the main text. Also listed are the transformation formulae between plane waves and spherical waves. Detailed discussion of these formulae may be found in a review article by Bostrom, Kristensson and Strom [15] as well as in References 5-7. The vector spherical wave functions are

$$\vec{\psi}_{1\ell m\sigma}(\vec{r}) = \left(\frac{k_p}{k_s}\right)^{1/2} \nabla \left[h_\ell^{(1)} (k_p r) \, Y_{\ell m\sigma} (\theta,\phi)\right] \tag{A.1}$$

$$\vec{\psi}_{2\ell m\sigma}(\vec{r}) = k_s \, \eta_\ell \, \nabla x \left[\vec{r} \, h_\ell^{(1)} (k_s r) \, Y_{\ell m\sigma}(\theta,\phi)\right] \tag{A.2}$$

$$\vec{\psi}_{3\ell m\sigma}(\vec{r}) = \frac{1}{k_s} \nabla \; x \; \vec{\psi}_{2\ell m\sigma} \tag{A.3}$$

where

$$\eta_\ell = [\ell(\ell+1)]^{1/2} \tag{A.4}$$

$$Y_{\ell m\sigma} = [\varepsilon_m \frac{(2\ell+1)(\ell-m)!}{4\pi(\ell+m)!}]^{1/2} \; P_\ell^m (\cos\theta) \begin{matrix} \cos m\phi \\ \\ \sin m\phi \end{matrix} \tag{A.5}$$

Equation A.1 represents an irrotational wave function and Equations A.2 and A.3 are solenoidal wave functions; $h_\ell^{(1)}$ are spherical Hankel functions of the first kind, $k_p$ and $k_s$ are the wavenumbers of longitudinal and transverse waves, $Y_{\ell m\sigma}$ are the spherical harmonics as defined in Equation A.5. The integer $\ell \in [0,\infty]$ and $m \in [0,\ell]$ and $\sigma$ denotes the even or odd parity of the $\phi$ dependence. Equations A.1-A.3 are singular at the origin. Solutions that are regular at the origin denoted by $\mathrm{Re}\psi$ may be obtained by replacing $h^{(1)}$ by $j_\ell$ the spherical Bessel functions. For notational convenience, we replace the subscripts on $\psi$ by the multiindex $n = \tau,\ell,m$ where $\tau = 1,2,3$ denotes the three possible polarizations of a vector field.

The vector plane waves are defined as

$$\vec{\phi}_1(\hat{\gamma},\vec{r}) = \frac{\hat{\gamma}}{4\pi}\left(\frac{k_p}{k_s}\right)^{1/2} \exp(ik_p\,\hat{\gamma}\cdot\vec{r}) \tag{A.6}$$

$$\vec{\phi}_2(\hat{\gamma},\vec{r}) = \frac{\hat{\alpha}}{4\pi}\exp(ik_s\hat{\gamma}\cdot\vec{r}) \tag{A.7}$$

$$\vec{\phi}_3(\hat{\gamma},\vec{r}) = \frac{\hat{\beta}}{4\pi}\exp(ik_s\hat{\gamma}\cdot\vec{r}) \tag{A.8}$$

where $\hat{\gamma},\hat{\alpha},\hat{\beta}$ analogous to $\hat{r},\hat{\theta},\hat{\phi}$ are unit vectors of a spherical coordinate system and $\hat{\gamma}$ has the Cartesian components

$$\hat{\gamma} = [\sin\alpha\cos\beta,\ \sin\alpha\sin\beta,\ \cos\alpha]. \tag{A.9}$$

However, though $\beta$ belongs to the real interval $(0,2\pi)$, $\alpha = \alpha'+i\alpha''$ where $0 \leq \alpha' \leq \pi$, $-\infty \leq \alpha'' \leq +\infty$. The wave functions $\phi_j^\dagger(\hat{\gamma},\vec{r})$ may be obtained from $\phi_j(\hat{\gamma},\vec{r})$ by conjugating the 'i' in the exponent.

The transformation formulae between plane waves and spherical waves are

$$\vec{\phi}_j(\hat{\gamma},\vec{r}) = -\frac{i}{k_s}\sum_n B_{nj}^\dagger(\hat{\gamma})\ \mathrm{Re}\,\vec{\psi}_n(\vec{r}) \tag{A.10}$$

$$\mathrm{Re}\,\vec{\psi}_n(\vec{r}) = i\,k_s\sum_j \int_0^\pi d\hat{\gamma}\ B_{nj}(\hat{\gamma})\ \vec{\phi}_j(\hat{\gamma},\vec{r}) \tag{A.11}$$

$$\vec{\psi}_n(\vec{r}) = 2i\, k_S \sum_j \int_{\substack{C_+;\, z>0 \\ C_-;\, z<0}} d\hat{\gamma}\, B_{nj}(\hat{\gamma})\, \vec{\phi}_j(\hat{\gamma},\vec{r}) \tag{A.12}$$

where for notational convenience we have used

$$\int_{C_\pm} d\hat{\gamma} = \int_{C_\pm} \sin\alpha\, d\alpha \int_0^{2\pi} d\beta$$

The integration contours $C_+$ and $C_-$ be in the complex $\alpha$ domain, $C_+$ is used if $\hat{z}.\vec{r} > 0$ and $C_-$ is used if $\hat{z}.\vec{r} < 0$, see Figure 3. The functions $B_{nj}(\hat{\gamma})$ are related to the vector spherical harmonics by the formulae

$$B_{nj}(\hat{\gamma}) = B_{\tau\ell m\sigma,j}(\hat{\gamma}) = (-1)^{-\ell-\delta_{\tau 3}}\, \hat{a}_j.\vec{A}_n(\hat{\gamma}) \tag{A.13}$$

where $\hat{a}_1 = \hat{\gamma}$, $\hat{a}_2 = \hat{\alpha}$ and $\hat{a}_3 = \hat{\beta}$ and

$$\vec{A}_{1\ell m\sigma}(\hat{\gamma}) = \hat{\gamma}\, Y_{\ell m\sigma}(\alpha,\beta) \tag{A.14}$$

$$\vec{A}_{2\ell m\sigma}(\hat{\gamma}) = \eta^{\ell}\, r\, \nabla\, Y_{\ell m\sigma}(\hat{\gamma}) \tag{A.15}$$

and

$$\vec{A}_{3\ell m\sigma}(\hat{\gamma}) = -\hat{\gamma} \times \vec{A}_{2\ell m\sigma}(\hat{\gamma}) \;. \tag{A.16}$$

Due to the properties of the vector spherical harmonics, we can show that there is no coupling of the longitudinal and transverse wave functions.

Since the formalism given in the main text is applicable to acoustic electromagnetic and elastic wave scattering, it is useful to describe which of the polarization indices $\tau$ and j contribute to each case. For the acoustic case, only longitudinal waves are present and hence $\tau = j = 1$ only. Further, the factor $\sqrt{k_p}/k_S$ in Equations A.1 and A.6 must be omitted and the factor $k_S$ in Equations A.10-A.12 must be replaced by $k_p$. For a transverse electromagnetic field $\tau = j = 2,3$ only. For an elastic field which admits both longitudinal (P-) and transverse (S-) waves propagating at distinct speeds, all the Equations A.1-A.16 must be retained as presented.

## APPENDIX B

### Translation of Scalar Spherical Functions

The scalar spherical functions that are singular at the origin, in accordance with Equation A.1 are given by

$$\psi_n(\vec{r}) = \psi_{\ell m\sigma}(\vec{r}) = h_\ell^{(1)}(kr)\, Y_{\ell m\sigma}(\theta,\phi) \; . \tag{B.1}$$

The translation of these functions is achieved by means of

$$\psi_n(\vec{r}-\vec{d}) = \begin{cases} \sum_{n'} R_{nn'}(-\vec{d})\, \psi_{n'}(\vec{r}) \; ; \; |\vec{r}| > |\vec{d}| \\ \\ \sum_{n'} \sigma_{nn'}(-\vec{d})\, \mathrm{Re}\, \psi_{n'}(\vec{r}) \; ; \; |\vec{r}| < |\vec{d}| \end{cases} \tag{B.2}$$

and

$$\mathrm{Re}\, \psi_n(\vec{r}-\vec{d}) = \sum_{n'} R_{nn'}(-\vec{d})\, \mathrm{Re}\, \psi_{n'}(\vec{r}) \tag{B.3}$$

where

$$R_{nn'} = \sigma_{nn'}(h^{(1)} \to j) \tag{B.4}$$

i.e., the R-matrix is obtained by replacing Hankel functions in the definition of $\sigma$ by Bessel functions. Expressions for $\sigma$ are given in Reference 15 and are written below:

$$\sigma_{nn'}(\vec{d}) = \sigma_{\ell m\sigma,\ell'm'\sigma}(\vec{d}) = (-1)^{m'} B_{\ell m,\ell'm'}(d,\eta) \cos(m-m')\psi$$

$$+ (-1)^{\sigma} B_{\ell m,\ell'-m'}(d,\eta) \cos(m+m')\psi \text{ if } \sigma = \sigma' \tag{B.5}$$

$$\sigma_{nn'}(\vec{d}) = \sigma_{\ell m\sigma,\ell'm'\sigma'} = (-1)^{m'+\sigma'} B_{\ell m,\ell'm'}(d,\eta) \sin(m-m')\psi$$

$$+ B_{\ell m,\ell'-m'}(d,\eta) \sin(m+m')\psi \text{ if } \sigma \neq \sigma' \tag{B.6}$$

where

$$B_{\ell m,\ell'm'}(d,\eta) = (-1)^{m+m'} \left(\frac{\varepsilon_m \varepsilon_{m'}}{4}\right)^{1/2} \sum_{\lambda=|\ell-\ell'|}^{\ell+\ell'} (-1)^{(\ell'-\ell+\lambda)/2} (2\lambda+1)$$

$$\left[\frac{(2\ell+1)(2\ell'+1)[\lambda-(m-m')]!}{(\lambda+[m-m'])!}\right]^{1/2} \begin{bmatrix} \ell & \ell' & \lambda \\ 0 & 0 & 0 \end{bmatrix} \begin{bmatrix} \ell & \ell' & \lambda \\ m & -m' & m'-m \end{bmatrix}$$

$$h_\lambda^{(1)}(kd)\, P_\lambda^{m-m'}(\cos\eta) \tag{B.7}$$

In the above equations, $\eta$ and $\psi$ are the spherical polar angles of the vector $\vec{d}$ and $\begin{bmatrix} \ell_1 & \ell_2 & \ell_3 \\ m_1 & m_2 & m_3 \end{bmatrix}$ are the Wigner 3-j symbols.

## REFERENCES


1. Datta, S. K. "Diffraction of SH-Waves by an Edge Crack," *J. Appl. Mech.* 46:101-106 (1979).

2. Datta, S. K. and El-Akily, N. "Diffraction of Elastic Waves by Cylindrical Cavity in a Half-Space," *J. Acoust. Soc. Am.* 64(6):1692-1699 (1978).

3. Stone, S. F., Ghosh, M. L., and Mal, A. K. "Diffraction of Antiplane Shear Waves by an Edge Crack," *J. Appl. Mech.* 47:359-362 (1980).

4. Mendelson, D. A., Achenbach, J. D. and Keer, L. M. "Scattering of Elastic Waves by a Surface-Breaking Crack," *Wave Motion* 2:277-292 (1980).

5. Kristenssön, G. and Ström, S. "Scattering from Buried Inhomogeneities - A General Three-Dimensional Formalism," *J. Acoust. Soc. Am.* 64(3):917-936 (1978).

6. Kristenssön, G. *J. Appl. Phys.* (In Press).

7. Boström, A. and Kristenssön, G. "Elastic Wave Scattering by a Three-Dimensional Inhomogeneity in An Elastic Half Space," *Wave Motion* 2:335-353 (1980).

8. Boström, A., Wall, D. J. N., Varadan, V. K. and Varadan, V. V., submitted for publication.

9. Varadan, V. V. and Varadan, V. K., "Configurations with Finite Numbers of Scatterers - A Self-Consistent T-Matrix Approach," *J. Acoust. Am.* 70(1):213-217 (1981).

10. Varadan, V. K. "Multiple scattering of acoustic, electromagnetic and elastic waves," in *Acoustic, Electromagnetic and Elastic Wave Scattering - Focus on the T-Matrix Approach*, V. K. Varadan and V. V. Varadan, Eds. (New York: Pergamon Press, 1980).

11. Waterman, P. C. "New Formulation of Acoustic Scattering," *J. Acoust. Soc. Am.* 45(6):1417-1429 (1969).

12. Waterman, P. C. "Symmetry, Unitarity, and Geometry in Electromagnetic Scattering," *Phys. Rev. D.* 3(4):825-839 (1971).

13. Waterman, P. C. "Matrix Theory of Elastic Wave Scattering," *J. Acoust. Soc. Am.* 60(3):567-580 (1976).

14. Varatharajulu, V. (Varadan, V. V.) and Pao, Y. H. "Scattering Matrix for Elastic Waves, I. Theory," *J. Acoust. Soc. Am.* 60(3):556-566 (1976).

15. Boström, A., Kristenssön, G. and Ström, S. "Transformation properties of plane, spherical and cylindrical scalar and vector wave functions - a review," to be published.

CHAPTER 6

# A DYNAMICAL EFFECTIVE MODULUS THEORY AND EXTENSIONS

*John J. McCoy*
The Catholic University of America
Washington, D.C. 20064

## ABSTRACT

A number of questions relating to the nature and the validity of a dynamical effective modulus theory and extensions are discussed within the framework of a general projection formalism. The relations of dynamical and statical effective property measures and the dependence of these measures on suitably chosen descriptors of the microscale geometry is considered.

## INTRODUCTION

As evidenced by this special symposium there has been a recent upsurge of interest in calculating the "average" response of a solid with a substructure, subjected to a dynamic loading. Although there are undoubtedly a fair number of specific applications prompting this interest, the two that have been brought to my attention are the effectiveness of acoustic coatings and acoustic propagation through porous media. There was a similar period of interest in the mid to late sixties, and early seventies. Then, the specific application prompting the interest was the response of composite materials to dynamic loadings. In reading the literatures of the studies reported in the two periods, one distinguishing feature would appear to be that the earlier work emphasized a continuum approach, a deterministic analysis, and periodic substructures whereas the more recent work emphasizes a multiple scatter approach, a statistical analysis, and irregular substructures. There were, of course, a number of studies reported in the sixties, and earlier, which interpreted the underlying physics as a multiple scattering and some conclusions reached in those studies are of interest. In my lecture and paper I will discuss these conclusions and point out what are, in my opinion, the more interesting of the remaining problems and tasks.

To set the stage for specific questions, we note that in a statistical interpretation of the fundamental problem, our interest is limited to the ensemble averaged response, or, as it is often termed in the scattering literature, the coherent field. This immediately restricts attention to experiments in which the signal wavelengths are large relative to the characteristic dimensions of the substructure. The reasons for this are two-fold. First, our primary interest is in sub-substructures that could be termed strongly inhomogeneous. For such substructures, the rate at which energy is scattered out of the coherent field is so rapid, when signal wavelengths are of the order of substructure dimensions, that there is little reason to discuss this component of the field further. Secondly, even for weakly inhomogeneous substructures if the signal wavelengths are small relative to substructure dimensions the most interesting information is in the fluctuating component of the propagating field, not in the coherent component. This latter situation describes the basic experiments when considering the scattering of electromagnetic signals by atmospheric turbulence, or, acoustic signals by oceanic internal waves. Each of these applications have resulted in extensive literatures which are of interest to the coherent field problem only because of a similarity of some mathematical techniques.

The validity of a dynamical effective modulus theory (DEMT) is expected to be asymptotic in the limits of long wavelength and low frequency. A DEMT does not predict long wavelength-high frequency solutions; i.e., its spectrum does not contain "optical" branches. A fundamental question that can be asked of a multiple scatter calculation, then, is the conditions, if any, under which one can expect coherent wave spectra with optical branches. The first questions to be asked, however, pertain to the long wavelength-low frequency limit. Can a DEMT be justified in this limit? Further, since it is unlikely that the predictions obtained using a DEMT, of the results of a propagation experiment for example, will be uniformly valid in range, a concomitant question is of the nature of the limit, or of estimates of propagation ranges in which observable errors can be expected. Finally, accepting the validity of a DEMT, the formulation of algorithms that determine the property measures required by the theory; i.e. the effective property measures; in terms of suitably defined measures of the substructure geometry can be addressed. In this last regard, it is natural to inquire as to the relationship between the property measures of the DEMT; i.e., the dynamical effective property measures; and analogous property measures of a statical effective modulus theory. It is in terms of these specific questions which I will frame my discussions.

## AN OPERATOR FORMULATION OF MEAN FIELD THEORIES

Consider the following fundamental problem. Accepting the validity of a heterogeneous, continuum formulation to be governing on one scale of observation; say a microscale; formulate a theory suitable for predicting the "average" response, which is observable on a larger scale of observation; say a macroscale. This problem has a long history, having received attention as far back as the early nineteenth century [1], and an extensive literature exists. Two classes of problems with different physical descriptions have come to serve as focal points of much of this literature. Continuum problems that adopt a deterministic and periodic modelling of the microstructure are approached by a procedure termed homogenization [2]. Continuum problems that adopt a statistical description of the microstructure are approached by a procedure termed a smoothing [3]. While reflecting different perspectives, the two procedures have evolved, with refinements in interpretation, to the point at which a common framework can be discerned [4]. This framework is contained in a projection method that is best outlined using an operator notation. Thus, consider the operator equation,

$$Lu = f , \qquad (1)$$

in which L is a linear operator with variations occurring on the microscale; f, a forcing with variations restricted to the macroscale; and u, the solution field. We introduce a projection operator, P; the precise definition of which differs in the two procedures. Ultimately, a projection operation is intended to result in a quantity that varies only on the macroscale. With P, the solution field can be resolved into a projected (or macro) field component, $\langle u \rangle = Pu$ and a fluctuating (or micro) field component, $u' = (I - P)u$, according to the identity

$$u = \langle u \rangle + u' . \qquad (2)$$

The identity operator is denoted by I.

Coupled equations can be written on the separate components by first projecting Equation 1, and by subsequently subtracting the projected equation from the original. We write

$$\langle L \rangle \langle u \rangle + \langle L'u' \rangle = f \qquad (3)$$

and

$$\langle L \rangle u' + (I - P) L'u'' = -L'\langle u \rangle . \qquad (4)$$

The essence of the projection method is to formally solve Equation 4 for the fluctuating field response, u', as a functional of the projected field response, <u>, and to substitute this formal solution into Equation 3. The result is a closed "macroscopic" equation of the form

$$(\langle L \rangle - M) \langle u \rangle = f, \tag{5}$$

in which the effective index operator (or, as it is alternately termed for historical reasons the mass operator), M, is expressed as

$$M = \langle L' [\langle L \rangle + (I - P) L' ]^{-1} L' \rangle \tag{6}$$

with $[\,]^{-1}$ denoting an inverse operator. An inverse operator must properly incorporate all required auxiliary conditions.

As discussed in [4], Equation 5 can provide a formal identity of homogenization and smoothing. The two approaches are distinguished from one another in two ways: as already noted, in the interpretation of the projection operator and, further, in the manner of effecting the solution of Equation 4. With respect to the projection operator, we note that while the precise definition of projection differs somewhat, reflecting the different perspectives of the microstructure, ultimately in both procedures projection is associated with an experimentally accessible volume average.

As outlined, the projection method leads to a calculational procedure for determining <u> which is carried out in two stages. First one calculates the effective index operator, M; i.e. one solves the microstructure equation; which, then, defines the macrostructure equation that is to be solved subsequently. An important question to now ask refers to accepting Equation 5 as the basis for a phenomenological model which governs the averaged response field. That is, can we change our interpretation of Equation 5 from a mathematically obtained partial integration of Equation 1 to a mathematically motivated "theory" that governs the macroscale response? In this latter interpretation the effective index operator is interpreted as a theory parameter, in the nature of a material property. Accepting this interpretation, the solution of Equation 4 is not accomplished analytically; instead, the effective index operator is to be determined in a (limited) number of tests. Although a number of mathematically subtle questions can be raised, the number depending on the degree of rigor demanded, accepting Equation 5 as the basis of a theory appears reasonable, for two widely separated length scales and for the linear formulations of interest here.

## APPLICATION OF FORMULATION TO PROPAGATION PROBLEMS

There were a number of studies reported throughout the sixties and early seventies in which the above described formulation was applied to the Helmholtz equation with a wavenumber that varied randomly with position [3,5-8], i.e.

$$(\nabla^2 + k_o^2\,[1 + \varepsilon\mu(\underset{\sim}{x})])\ \phi = 0 \tag{7}$$

Here $\varepsilon$ provides a measure of the strength of the variations in the refractive index field and $\mu(x)$ is a centered (i.e. $\langle\mu\rangle \equiv 0$) random function of position. The effective index operator for the macroscale equation on the mean response field, $\langle\phi(x)\rangle$, is a nonlocal refractive index field, i.e.

$$(\nabla^2 + k_o^2)\langle\phi(\underset{\sim}{x})\rangle + \int\ k(\underset{\sim}{x},\underset{\sim}{x}')\langle\phi(\underset{\sim}{x}')\rangle d\underset{\sim}{x}' = 0 \tag{8}$$

For an unbounded medium and homogeneous statistics, the integral in Equation 8 is a convolution enabling us to readily obtain a characteristic equation that determines the plane wave solutions. The most commonly reported calculations assume that $\varepsilon \ll 1$, which enables a perturbation solution of the microstructure equation, thereby obtaining an analytic expression for $K(\underset{\sim}{x} - \underset{\sim}{x}')$. To lowest order, this expression depends on the two-point correlation function of the randomly varying $\mu(\underset{\sim}{x})$ field. Accepting certain reasonable properties for this correlation function a number of conclusions of the mean field dispersion spectrum can be demonstrated. Among these are that the spectrum is dispersive and that all plane waves will exhibit decay, for realistic random medium. Further, the spectrum contains a multiplicity of branches with a single branch in the low frequency-long wavelength part of the spectrum. This branch is the counterpart of the single branch that is obtained for the medium without a microstructure. The effect of the microstructure, on this branch, is to introduce dispersion and dissipation. In [8], the author considered the possibility of high frequency-long wavelength plane wave solutions; i.e., optical branches; and demonstrated these do not exist, for the calculation accomplished. This appears reasonable since one can trace the existence of optical branches to constructive interference of coherently scattered signals. It is the lack of coherence of randomly scattered signals that results in both the perceived dissipation in the acoustic branch and the loss of propagating optical branches. While the additional branches in the dispersion spectrum of a medium with a microstructure do not correspond to propagating waves, they would result in boundary layers. The possible importance of these layers on reflection coefficients from a medium with a microstructure has not, to my knowledge, been answered.

The general formulation has also been applied to the dynamical equations of elasticity [9,10], i.e.

$$\nabla \cdot \underset{\sim}{\tau} + \rho\omega^2 \underset{\sim}{u} = \underset{\sim}{f} ,$$

$$\underset{\sim}{\tau} = \underset{\sim}{C}:\underset{\sim}{\varepsilon} , \tag{9}$$

and

$$\underset{\sim}{\varepsilon} = \frac{1}{2}(\nabla\underset{\sim}{u} + \underset{\sim}{u}\nabla) .$$

Here both the mass density, $\rho(x)$, and the material moduli tensor $\underset{\sim}{C}(\underset{\sim}{x})$ are described by stochastic functions of position. (The detailed calculations are usually accomplished for an isotropic material moduli tensor.) The general formulation is quite complex. In [10], the author studied approximations to the general formulation, which are suitable for reproducing solutions in the low frequency-long wavelength portion of the plane wave dispersion spectrum. Several conclusions were demonstrated. First, was the validity of a DEMT, as a zero$^{th}$ order theory, in which it is proper to use an "average" mass density and statical values for the effective material moduli. The physical reason behind this conclusion is made clear by the analysis leading to it. The reason is that the microscale problem; i.e., that represented by Equation 4; remains a statical problem in the long wavelength-low frequency limit. That is, while the theory is suitable for experiments that appear dynamical for lengths on the macroscale, the experiments appear statical for lengths on the microscale. Further, in [10], corrections to the low frequency-long wavelength, effective modulus theory were shown to contain both dispersion and dissipation effects. The importance of these effects can be estimated by the propagation range required for them to be measurable. Although there is a degree of arbitrariness in the definition of measurable, it is possible to provide order of magnitude range estimates. The estimates obtained in [10], which appear to depend on the dimensionality of the problem, were $L^3/\ell^2$ for dispersion and $L^4/\ell^3$ for dissipation, where L is the signal wavelength and $\ell$ is a measure of the microscale length.

## ALTERNATE CALCULATIONS OF EFFECTIVE PROPERTY MEASURES

As presented in Section II, the fundamental problem motivating the described formulations concentrated on the derivation of macroscale field equations which, under some circumstances, can be properly interpreted to provide the basis of a phenomenological theory. The identification of effective property measures is achieved quite naturally, as a by-product in such an approach. Alternate definitions and calculations of effective property measures are based on matching solutions in specific physical, or thought, experiments. A possibility of confusion can arise as a result of different investigators using identical terminology to describe different entities.

Statical effective moduli are most commonly defined for specimens required to be large enough, compared to all microscale lengths, to be representative and loaded in such a manner that the macroscale stress and strain fields are homogeneous; i.e., are described by constant tensors. The effective moduli are, then, most naturally defined as the ratios of components of these tensors. Or, in what can be shown to be an equivalent definition, the effective moduli can be defined on equating the energy density in the effective medium to the averaged energy density of the material with microstructure. It has been shown [11] that these physically understood definitions of the effective moduli are consistent with the mathematical identification given by the general formulation. The value of the alternate definitions lie in the fact that they describe a physical experiment for measuring them and, especially in the case of the energy definition; provide a framework for relating the values of the effective property measures to geometric descriptions of the microstructure.

Dynamical effective property measures have been defined in terms of a number of propagation/scattering, thought experiments. A common calculation, especially for a medium that consists of a suspension of a particulate phase of one material dispersed throughout a matrix phase of a second material, is the effective wave speed for a propagating mean, or macroscale, plane wave [12]. A less common calculation, but one that has received recent attention [13], matches aspects of the scattering crossection of a heterogeneous scattering volume to that of an effective homogeneous scattering volume. There has been no attempt to relate the dynamical effective property measures defined as above to a general formulation such as that outlined in Section II, although the long wavelength-low frequency limit is sometimes related to the large literature of statical effective moduli studies. The advantage claimed for the definitions of the dynamical effective measure definitions, besides the generalization to incorporate time varying phenomenon, is that they suggest a number of approximations that have been used in the literature devoted to scattering theories.

## EFFECTIVE PROPERTY PREDICTION MODELS

The relationship between effective property measures and suitably chosen descriptors of the microscale geometry was one of the primary questions prompting composite materials research and remains largely unanswered. The question has interested me since it seems clear that numerical values of effective property measures can be quite sensitive to refined information of the microscale geometry. For example, it is well appreciated that inclusion shape can greatly influence the effective properties of a suspension, even at low values of concentration. It is also true, although apparently less appreciated, that introducing the possibility of any preferred clustering of inclusions can likewise influence effective property measures of a suspension. Even the use of a single-body calculation to provide a first-order; i.e. O(c) - c being the concentration; estimate of an effective property of a dilute suspension assumes that no clustering can occur [14]. Further, it is intuitive that void connectivity measures should greatly influence effective properties of porous materials. It would seem to be natural, then, for a materials scientist with an interest in designing a composite material with "optimum" effective properties, or, for a physicist interested in an inverse problem in which the microscale geometry is to be inferred from effective property measures, to exploit this sensitivity. Thus, it has been a source of some surprise that more effort is spent on developing models that appear to assume from the outset that effective property measures are largely insensitive to refined information of the microscale geometry. The fact that a self-consistent calculation might provide "reasonable answers" for a particular material manufactured in a particular manner could be of interest, if one is confined to "similar" materials manufactured in a "similar" manner. A far more interesting question, however, would appear to consider the possibility of altering the manufacturing process in such a way as to alter a value of an effective property measure by altering the microscale geometry. A self-consistent calculation cannot address this question since it assumes; in general, falsely; that the value of an effective property measure does not depend on refined information of the microscale geometry. The inverse problem provides an even more obvious illustration of an application in which sensitivity of an effective property measure to refined information of microscale geometry is to be exploited, if possible. Acoustic logging data providing estimates of soil porosity may be of interest but prediction models that would enable estimates of measures of void connectivity from logging data would appear to be of greater interest. A discussion of some research on relating effective property measures to microscale geometry is provided in [15], along with original references.

ACKNOWLEDGMENT

This paper and a study on the importance of higher-order information of microscale geometry on effective property measures is supported by a National Science Foundation Grant (ENG-7820613). This support is acknowledged as is the continuing encouragement and support of Dr. Clifford Astil of NSF.

## REFERENCES


1. Poisson, S. D. "Second mem sur la Theorie de Magnetisme, *Mem. L'Acad. France* 5 (1822).

2. Babuska, I. In: *Numerical Solution of Partial Differential Equations III*, B. Hubbard, Ed. (New York: Academic Press, 1976), p. 89.

3. Frisch, U. In: *Probabilistic Methods in Applied Mathematics I*, A. T. Bharucha-Reid, Ed. (New York: Academic Press, 1968), p. 76.

4. Fishman, L. and McCoy, J. J. "A Unified View of Bulk Property Theories for Stochastic and Periodic Media," *ZAMP* 32(1):45-61 (1981).

5. Bourret, R. C. "Propagation of Randomly Perturbed Fields," *Canad. J. Phys.* 40(6):782-790 (1962).

6. Keller, J. B. In: *Proc. Symp. Applied Mathematics*, G. Birkhoff, R. Bellman and C. C. Lin, Eds. (Providence, RI: American Math. Soc., 1962), p. 227.

7. Tartarski, V. I. and Gertsenshtein, M. E. "Propagation of Waves in a Medium with Strong Fluctuations of the Refractive Index," *Zh. Eksperim. Teoret. Fiz.* 44(2):676-685 (1963).

8. McCoy, J. J. "A Note on the Mean Field Dispersion Spectra for Media with Disordered Substructures," *J. Appl. Mech.* 39(4):1155-1156 (1972).

9. Karal, F. and Keller, J. B. "Elastic, Electromagnetic and Other Waves in a Random Medium," *J. Math. Phys.* 5(4):537-547 (1964).

10. McCoy, J. J. "On the Dynamic Response of Disordered Composites," *J. Appl. Mech.* 40(2):511-517 (1973).

11. McCoy, J. J. "On the Calculation of Bulk Properties of Heterogeneous Materials," *Q. Appl. Math.* 37(2):137-149 (1979).

12. See, for example, the articles by U. Frisch [3].

13. See other papers of this symposium for a representative sampling of references.

14. McCoy, J. J. and Beran, M. J. "On the Thermal Conductivity of a Random Suspension of Spheres," *Int. J. Engr. Sci.* 14(1):7-18 (1976).

15. McCoy, J. J. In: *Mechanics Today, Vol. 6*, S. Nemat-Nasser, Ed. (New York: Pergamon Press, 1981), p. 1. See also other papers from this symposium.

CHAPTER 7

# EFFECTIVE MEDIUM THEORY FOR ELASTIC COMPOSITES

*James G. Berryman*
Bell Laboratories
Whippany, NJ

The theoretical foundations of a variant on effective medium theories for elastic constants of composites are discussed. The connection between this approach and the methods of Zeller and Dederichs, Korringa, and Gubernatis and Krumhansl is elucidated. A review of the known relationships between the various effective medium theories and rigorous bounding methods for elastic constants is also provided.

In recent series of papers [1-3] a variant on effective medium theories for elastic composites has been developed by the present author. In this paper, we will review the derivation of the effective medium formulas for the elastic constants of composites and we will elucidate the relationship between our results and the results of other effective medium theories. We also compare our results to known rigorous bounds on the effective elastic constants.

The general background for theories of elastic composites is discussed in the recent review article by Watt, Davies, and O'Connell [4]. A review of effective medium theories is given by Elliott, Krumhansl, and Leath [5].

## EFFECTIVE ELASTIC CONSTANTS

Mal and Knopoff [6] derived an integral equation for the scattered displacement field from a single elastic scatterer. Let $\Omega_i$ be the volume of the region occupied by a single inclusion. Let the incident field be $\vec{u}^0(\vec{x})\exp(-i\omega t)$ and let $\vec{u}(\vec{x})\exp(-i\omega t)$ and $\vec{v}(\vec{x})\exp(-i\omega t)$ be the total field outside and inside the inclusion volume such that

$$\vec{u}(\vec{x}) = \vec{u}^0(\vec{x}) + \vec{u}^s(\vec{x}) \quad \text{for} \quad \vec{x} \notin \Omega_i$$

and (1)

$$\vec{v}(\vec{x}) = \vec{u}^0(\vec{x}) + \vec{v}^s(\vec{x}) \quad \text{for} \quad \vec{x} \in \Omega_i \ .$$

The scattered fields are $\vec{u}^s$ and $\vec{v}^s$. Both $\vec{u}(\vec{x})$ and $\vec{u}^0(\vec{x})$ satisfy the same equation

$$c^m_{\ell npq} \frac{\partial^2 u_p}{\partial x_n \partial x_q} + \rho_m \omega^2 u_\ell = 0 \tag{2}$$

while $\vec{v}(\vec{x})$ satisfies

$$c^i_{\ell npq} \frac{\partial^2 v_p}{\partial x_n \partial x_q} + \rho_i \omega^2 v_\ell = 0 \tag{3}$$

The indices $\ell,n,p,q$ take the values 1,2,3, and the summation convention applies in Equations 2 and 3 and throughout the paper. The elastic tensor for the matrix and inclusion are respectively

$$c^m_{\ell npq} = \lambda_m \delta_{\ell n} \delta_{pq} + \mu_m (\delta_{\ell p} \delta_{nq} + \delta_{np} \delta_{\ell q}) \tag{4}$$

$$c^i_{\ell npq} = \lambda_i \delta_{\ell n} \delta_{pq} + \mu_i (\delta_{\ell p} \delta_{nq} + \delta_{np} \delta_{\ell q})$$

$$\equiv c^m_{\ell npq} + \Delta c^i_{\ell npq} , \tag{5}$$

and $\rho_m$, $\rho_i$ $(\equiv \rho_m + \Delta\rho^i)$ are the respective densities.

Green's function for a point source in an infinite, isotropic, homogeneous elastic medium of the matrix material is given by

$$g_{pq}(\vec{x},\vec{\zeta}) = \frac{1}{4\pi\rho_m\omega^2} \left[ s^2 \frac{e^{isr}}{r} \delta_{pq} - \frac{\partial^2}{\partial x_p \partial x_q} \left( \frac{e^{ikr}}{r} - \frac{e^{isr}}{r} \right) \right] \tag{6}$$

where $r = |\vec{x}-\vec{\zeta}|$, $k = \omega[\rho_m/(\lambda_m+2\mu_m)]^{1/2}$, $s = \omega[\rho_m/\mu_m]^{1/2}$ -- $k$ and $s$ being, respectively, the magnitudes of the wave-vectors for compressional and shear waves in the matrix. Given $g_{pq}$, Mal and Knopoff [6] then derive an integral equation for $\vec{u}(\vec{x})$. Since the derivation follows standard lines of argument [3], we need not repeat it here. The result is

$$u_\ell(\vec{x}) = u^0_\ell(\vec{x}) + \int_{\Omega_i} d\vec{\zeta} \left[ \Delta\rho^i \omega^2 v_n(\vec{\zeta}) - \Delta c^i_{njpq} \varepsilon_{pq} \frac{\partial}{\partial \zeta_j} \right] g_{\ell n}(\vec{x},\vec{\zeta}). \tag{7}$$

Equation 7 is an exact integral equation for the displacement field in the region exterior to the scatterer in terms of the displacement and strain field inside $\Omega_i$.

To evaluate the integral (7), an estimate of the interior displacement and strain is required. Recalling the first Born approximation in quantum scattering theory suggests the estimates

$$\vec{v}(\vec{\zeta}) \simeq \vec{u}^0(\vec{\zeta}) \tag{8}$$

and for $\vec{\zeta} \in \Omega_i$.

$$\varepsilon_{pq}(\vec{\zeta}) = \varepsilon^0_{pq}(\vec{\zeta}) \tag{9}$$

By Equations 8 and 9 we mean to approximate $\vec{v}$ and $\varepsilon$ by the values $\vec{u}^0$ and $\varepsilon^0$ would have achieved at $\vec{\zeta}$ if the matrix contained no scatterers. For scatterers with small volumes, it follows from (7) that $\vec{u}^s(\vec{x})$ and its derivatives are small quantities for $\vec{x}$ outside of $\Omega_i$. Since the displacement is continuous across the boundary, we see that Equation 8 will be a good approximation to $\vec{v}(\vec{\zeta})$. However, this argument fails for Equation 9 because the strains are not continuous across the boundary. Equation 9 should therefore be replaced by the formula

$$\varepsilon_{pq} = T_{pqrs}\varepsilon^0_{rs} , \tag{10}$$

where T is Wu's tensor [7] relating $\varepsilon_{pq}$ for an arbitrary ellipsoidal inclusion to the uniform strain at infinity $\varepsilon_{pq}$. Now, if the wavelength of the incident waves is large compared to the size of the ellipsoid ($a/\lambda \ll 1$), then the field both near the ellipsoid and inside $\Omega_i$ will be essentially static and uniform [8]. Thus, to the lowest order of approximation, it is valid to make the substitutions (8) and (10). When the ellipsoid is centered at $\vec{\zeta}_i$, we find easily that

$$u^s_\ell(\vec{x}) = \Omega_i \Big\{ \Delta\rho^i \omega^2 u^0_n(\vec{\zeta}_i) g_{\ell n}(\vec{x},\vec{\zeta}_i) - \Big[ \Delta\lambda^i T_{pprs}\delta_{nj} + 2\Delta u^i T_{njrs} \Big] \varepsilon^0_{rs} g_{\ell n,j}(\vec{x},\vec{\zeta}_i) \Big\} \tag{11}$$

where the symmetry properties of T have been used in simplifying the expression. A comma preceding a subscript indicates a derivative.

Equation 11 gives the first order estimate of the scattered wave from an ellipsoidal inclusion whose principal axes are aligned with the coordinate axes. When the ellipsoid is oriented arbitrarily with respect to the coordinate axes,

Equation 11 must be changed by replacing $T_{pqrs}$ everywhere by

$$U_{pqrs} = \ell_{p\alpha}\ell_{q\beta}\ell_{r\gamma}\ell_{s\delta}T_{\alpha\beta\gamma\delta} \tag{12}$$

where $\ell_{\alpha\beta}$ are the appropriate direction cosines. For homogeneous, isotropic composites with randomly oriented ellipsoidal inclusions, the general form of the average tensor is [7]

$$\overline{U}_{pqrs} = \frac{1}{3}(P-Q)\delta_{pq}\delta_{rs} + \frac{1}{2}Q(\delta_{pr}\delta_{qs}+\delta_{ps}\delta_{qr}), \tag{13}$$

where

$$P = \frac{1}{3}\,T_{ppqq} \text{ and } Q = \frac{1}{5}(T_{pqpq} - \frac{1}{3}\,T_{ppqq}). \tag{14}$$

Finally, suppose N inclusions are contained in a small volume of radius a centered at $\vec{\zeta}_0$. Assume that the effects of multiple scattering may be neglected at sufficiently low frequencies to the lowest order. Then, to the same degree of approximation used in Equation 11 (i.e., $a/\lambda << 1$), the scattered wave has the form

$$\langle u_\ell^s(\vec{x})\rangle^m \cong \sum_{i=1}^{N} \Omega_i \{\Delta\rho^i\omega^2 u_n^0(\vec{\zeta}_0)g_{\ell n}(\vec{x},\vec{\zeta}_0) - [\Delta\lambda^i\overline{U}^{mi}_{pprs}\delta_{nj} + 2\Delta\mu^i\overline{U}^{mi}_{njrs}]\varepsilon^0_{rs}g_{\ell n,j}(\vec{x},\vec{\zeta}_0)\} \tag{15}$$

where the superscripts m and i refer to matrix and inclusion properties, respectively. Distinct superscripts i must be used in Equation 15 to specify both the material and the shape of the inclusion.

Now to obtain the effective medium estimates for the elastic constants, assume that a small sphere of the true composite material is imbedded in a medium whose elastic properties may be varied freely in a controlled manner so the properties of the imbedding medium are always known. Then, conceptually, one way to estimate the effective (low frequency) elastic constants of the composite is to do a series of long-wavelength scattering experiments while varying the properties of the imbedding medium until there is no scattering from the composite sphere. The elastic constants of the composite must then be identical to the elastic constants of the imbedding medium.

To apply this thought experiment to the analytical problem of estimating elastic constants, consider replacing the true composite sphere with a sphere composed of matrix material identical to the imbedding material and of ellipsoidal inclusions of the same materials as those in the true composite and in the same proportion. Then, if multiple scattering effects are neglected, the theoretical expression which determines the elastic constants is

$$\langle u_\ell^s(x) \rangle^* = 0 \tag{16}$$

where the left hand side is given by Equation 15 with matrix-type m=*. Equation 16 states that the net scattering in the self-consistently determined medium vanishes to lowest order. If the volume fraction is defined by $f_i = \Omega_i / \sum_{j=1}^{N} \Omega_j$, then Equation 16 implies the following formulas:

$$\sum_{i=1}^{N} f_i(\rho_i - \rho^*) = 0 \ , \tag{17}$$

$$\sum_{i=1}^{N} f_i(K_i - K^*)P^{*i} = 0 \ , \tag{18}$$

and

$$\sum_{i=1}^{N} f_i(\mu_i - \mu^*)Q^{*i} = 0 \ . \tag{19}$$

Equation 17 states that the effective density $\rho^*$ is just the average density. Equations 18 and 19 provide implicit formulas for $K^*$ and $\mu^*$. These implicit formulas may be solved numerically by iteration [3].

Equations 18 and 19 were obtained independently by Korringa, Brown, Thompson, and Runge [9] using an entirely different method. In the following sections, we will compare the results obtained from our effective medium theory to the known rigorous bounds on elastic constants and to the results of other effective medium theories.

## RIGOROUS BOUNDS ON EFFECTIVE MODULI

In their review, Watt, Davies, and O'Connell [4] discuss various rigorous bounds on the effective moduli of composites. For example, the well-known Voigt (arithmetic) and Reuss (harmonic) averages are respectively rigorous [10] upper and lower bounds for both $K^*$ and $\mu^*$. Generally tighter bounds have been given by Hashin and Shtrikman [11].

Still righter bounds have been obtained in principle by Beran and Molyneux [12] for the bulk modulus and by McCoy [13] for the shear modulus. However, the resulting formulas depend on 3-point correlation functions for the composite and are therefore considerably more difficult to evaluate than the expressions for the Hashin-Shtrikman bounds. Miller [14] evaluated the bounds of Beran and Molyneux by treating an isotropic homogeneous distribution of statistically independent cells. Silnutzer [15] used the same approach to simplify the bounds of McCoy for cell materials. Recently, Milton [16] has shown that the bounds of Beran and Molyneux and of McCoy can be simplified somewhat even if the composite is not a cell material. Nevertheless, the bounds which are most easily evaluated are the Hashin - Shtrikman bounds, the Beran - Molyneux - Miller bounds, and the McCoy - Silnutzer bounds. We will compare these bounds to the estimates obtained from the effective medium theory.

To aid in our comparisons, it is convenient to introduce the following functions

$$K(x) = \left( \sum_{i=1}^{N} \frac{f_i}{K_i + \frac{4}{3}x} \right)^{-1} - \frac{4}{3}x \, , \tag{20}$$

$$\mu(y) = \left( \sum_{i=1}^{N} \frac{f_i}{\mu_i + y} \right)^{-1} - y \, , \tag{21}$$

and

$$F(x,z) = \frac{x}{6}(9z + 8x)/(z + 2x) \; . \tag{22}$$

It has been shown previously [2] that $K(x)$ and $\mu(y)$ are monotonically increasing functions of their arguments. Similarly, we find

$$\frac{\partial F}{\partial z} = \frac{5x^2}{3(z+2x)^2} \geq 0 \tag{23}$$

and

$$\frac{\partial F}{\partial x} = (9z^2+16xz+16x^2)/6(z+2x)^2 \; . \tag{24}$$

When both arguments of $F(x,z)$ are positive, it follows that $F$ is a monotonically increasing function of both arguments.

Now, if we define the minimum and maximum moduli among all the constituents by

$$\mu_+ = \max(\mu_1,\ldots,\mu_N),\ \mu_- = \min(\mu_1,\ldots,\mu_N)$$

$$K_+ = \max(K_1,\ldots,K_N),\ K_- = \min(K_1,\ldots,K_N)\ , \tag{25}$$

then the Hashin - Shtrikman bounds are given in general by

$$K^{\pm}_{HS} = K(\mu_{\pm}) \tag{26}$$

and

$$\mu^{\pm}_{HS} = \mu[F(\mu_{\pm},\ K_{\pm})]\ . \tag{27}$$

The Beran-Molyneux-Miller bounds and the McCoy-Silnutzer bounds are known for 2-phase composites (i.e., N=2). These bounds can be written in concise form using the notation of Milton [16]. If we define two geometric parameters $\zeta_1 = 1-\zeta_2$ and $\eta_1 = 1-\eta_2$ and the three averages of any modulus M by $\langle M\rangle = M_1f_1 + M_2f_2$, $\langle M\rangle_\zeta = M_1\zeta_1 + M_2\zeta_2$ and $\langle M\rangle_\eta = M_1\eta_1 + M_2\eta_2$, then the bounds can be written as

$$K^{+}_{BMM} = K(\langle\mu\rangle_\zeta)\ , \tag{28}$$

$$K^{-}_{BMM} = K(\langle\frac{1}{\mu}\rangle_\zeta^{-1}), \tag{29}$$

$$\mu^{+}_{MS} = \mu(\theta/6), \tag{30}$$

and

$$\mu^{-}_{MS} = \mu(\Xi^{-1}/6) \tag{31}$$

where

$$\theta = [10\langle\mu\rangle^2\langle K\rangle_\zeta+5\langle\mu\rangle\langle 3\mu+2K\rangle\langle\mu\rangle_\zeta+\langle 3K+\mu\rangle^2\langle\mu\rangle_\eta]/\langle K+2\mu\rangle^2 \tag{32}$$

and

$$\Xi = [10\langle K\rangle^2\langle\frac{1}{K}\rangle_\zeta+5\langle\mu\rangle\langle 3\mu+2K\rangle\langle\frac{1}{\mu}\rangle_\zeta+\langle 3K+\mu\rangle^2\langle\frac{1}{\mu}\rangle_\eta]/\langle 9K+8\mu\rangle^2\ . \tag{33}$$

For symmetric cell materials, $\zeta_1 = \eta_1 = f_1$ for spherical cells, $\zeta_1 = \eta_1 = f_2$ for disks, and $\zeta_1 = (f_2+3f_1)/4$, $\eta_1 = (f_2+5f_1)/6$ for needles.

It is particularly simple to compare these bounds with the results of effective medium theory when the inclusions are assumed to be spherical in shape. Then, the estimates of the moduli are given by

$$K^* = K(\mu^*) \tag{34}$$

$$\mu^* = \mu[F(\mu^*, K^*)]. \tag{35}$$

Furthermore, the bounds (28-30) simplify in this case and are given by

$$K^+_{BMM} = K(\langle\mu\rangle), \tag{36}$$

$$K^-_{BMM} = K(\langle\frac{1}{\mu}\rangle^{-1}), \tag{37}$$

and

$$\mu^+_{MS} = \mu[F(\langle\mu\rangle, \langle K\rangle)]. \tag{38}$$

From the monotonicity properties of the functions (20-22) and from elementary arguments relating the estimates to the Voigt and Reuss averages, we find for the bulk modulus that

$$K(\mu_-) \leq K(\langle\frac{1}{\mu}\rangle^{-1}) \leq K(\mu^*) \leq K(\langle\mu\rangle) \leq K(\mu_+) \tag{39}$$

or equivalently

$$K^-_{HS} \leq K^-_{BMM} \leq K^* \leq K^+_{BMM} \leq K^+_{HS} . \tag{40}$$

Similarly, it follows for the shear modulus that

$$\mu^-_{HS} \leq \mu^* \leq \mu^+_{MS} \leq \mu^+_{HS} \tag{41}$$

The detailed argument leading to Equation 39 is a little involved: We must first show that $K^*$, $\mu^*$ are bounded by the Hashin - Shtrikman bounds [2]. Then, since the Hashin - Shtrikman bounds are themselves bounded by the Voigt and Reuss bounds, Equation 39 follows from

$$K(\langle\frac{1}{\mu}\rangle^{-1}) \leq K(\mu^-_{HS}) \leq K(\mu^*) \leq K(\mu^+_{HS}) \leq K(\langle\mu\rangle). \tag{42}$$

The arguments just given are valid only for the case of spherical inclusions. The author knows of no general argument relating the effective medium results to the rigorous bounds for arbitrary inclusion shapes. However, numerical examples have shown that the effective medium estimates always lie between the bounds.

Typical results are presented in Figures 1-3. The values of the constituent moduli were chosen to be: $K_1 = 0.44$ Mb, $\mu_1 = 0.37$ Mb, $K_2 = 0.14$ Mb, and $\mu_2 = 0.10$ Mb. The values of $K_2$ and $\mu_2$ were chosen as a compromise between two extremes: (1) If $K_2$, $\mu_2$ are too close to $K_1$, $\mu_1$, the bounds are too close together to be distinguishable on the plot. (2) If $K_2$, $\mu_2$ are both zero, the iteration to obtain the effective medium results does not converge for the case of disk inclusions [3]. We find in all cases considered that the effective medium theory results lie between the rigorous bounds.

## OTHER EFFECTIVE MEDIUM THEORIES

Various effective medium theories of the elastic properties of composites exist. Of these theories, the scattering theory presented by Zeller and Dederichs [17], Korringa [18], and Gubernatis and Krumhansl [19], has the most in common with our scattering-theory approach. However, our approach appears to be unique among the self-consistent scattering-theory variety, being dynamic while all others are based on static derivations. We hope to be able to take advantage of this uniqueness when we generalize this approach to finite frequencies.

Another class of effective medium theories studied by Hill [20], Budiansky [21], Wu [7], Walpole [22], and Boucher [23] does not yield the same results as our approach except for the case of spherical inclusions. We have shown elsewhere [3] how the derivation of this latter class of theories can be symmetrized to yield our results. Since this class of effective medium theories gives results equivalent to the Hashin - Shtrikman bounds when the inclusions are disk-shaped, we conclude that our results are preferred.

To elucidate the relationship between the static and dynamic derivations of the effective medium results, we will outline the static derivation. The integral equations for the static strain field is given by

$$\varepsilon_{ij}(\vec{x}) = \varepsilon^0_{ij}(\vec{x}) + \int d^3x' G_{ijk\ell}(\vec{x},\vec{x}')\Delta c_{k\ell mn}(\vec{x}')\varepsilon_{mn}(\vec{x}') \quad (43)$$

where

$$G_{ijk\ell}(\vec{x},\vec{x}') = \frac{1}{2}(g^0_{ik,j\ell}+g^0_{jk,i\ell}). \quad (44)$$

The Kelvin solution is

$$g^0_{pq}(\vec{x},\vec{x}') = \frac{1}{4\pi\mu_m}\left[\frac{\delta_{pq}}{r} - \frac{1}{4(1-\nu_m)}\frac{\partial^2 r}{\partial x_p \partial x_q}\right] \quad (45)$$

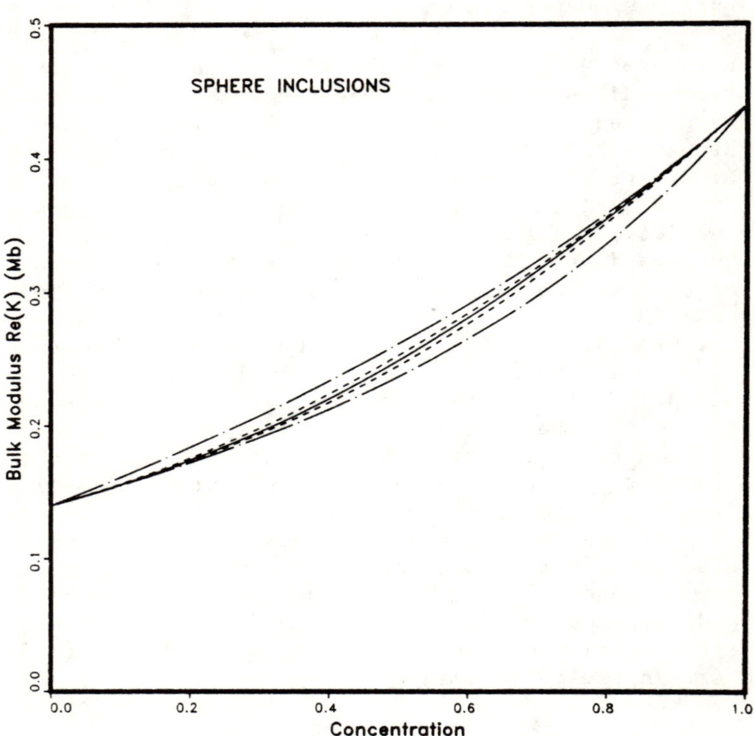


Figure 1(a). Estimates of the effective bulk modulus of elastic composites with constituents $K_1$=0.44 Mb, $\mu_1$=0.37 Mb, $K_2$=0.14 Mb, and $\mu_2$=0.10 as the concentration (volume fraction) of type-1 constituent increases. The curves are respectively the effective medium theory estimate (solid line), Beran - Molyneux - Miller bounds (dash lines), and Hashin - Shtrikman bounds (dot - dash lines). Inclusions are assumed spherical in shape.

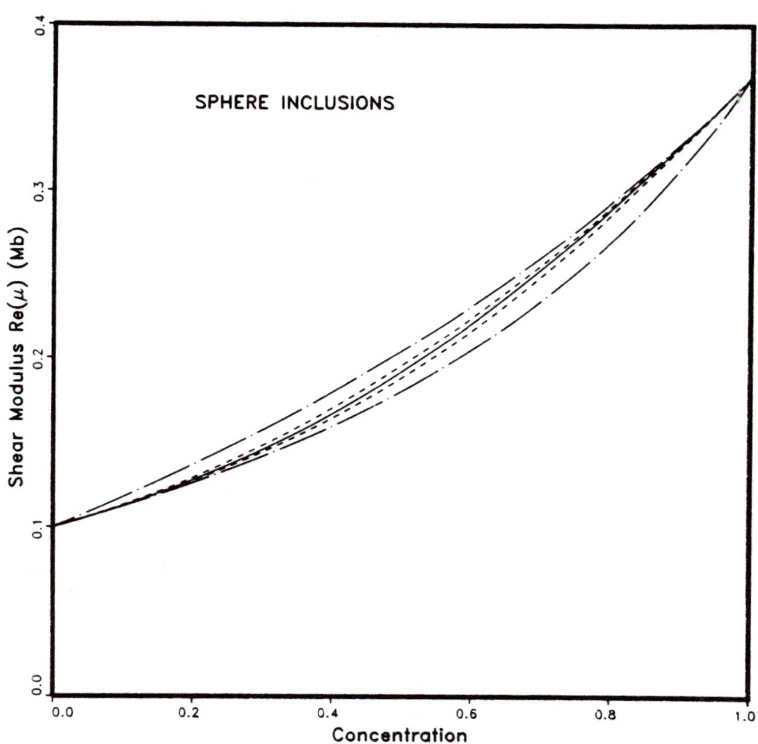


Figure 1(b). Estimates of the effective shear modulus of elastic composites with constituents as in Figure 1(a). The curves are respectively the effective medium theory estimate (solid line), McCoy - Silnutzer bounds (dash lines), and Hashin - Shtrikman bounds (dot - dash lines). Inclusions are assumed spherical in shape.

where $r = |\vec{x}-\vec{x}'|$ and $\mu_m$ and $\nu_m$ are the shear modulus and Poisson's ratio of the matrix. Equation 43 may be rewritten formally as

$$\varepsilon = \varepsilon^0 + G\Delta c\varepsilon \tag{46}$$

where G is now an integral operator defined by

$$Gf = \int d^3x' G(\vec{x},\vec{x}') f(\vec{x}'). \tag{47}$$

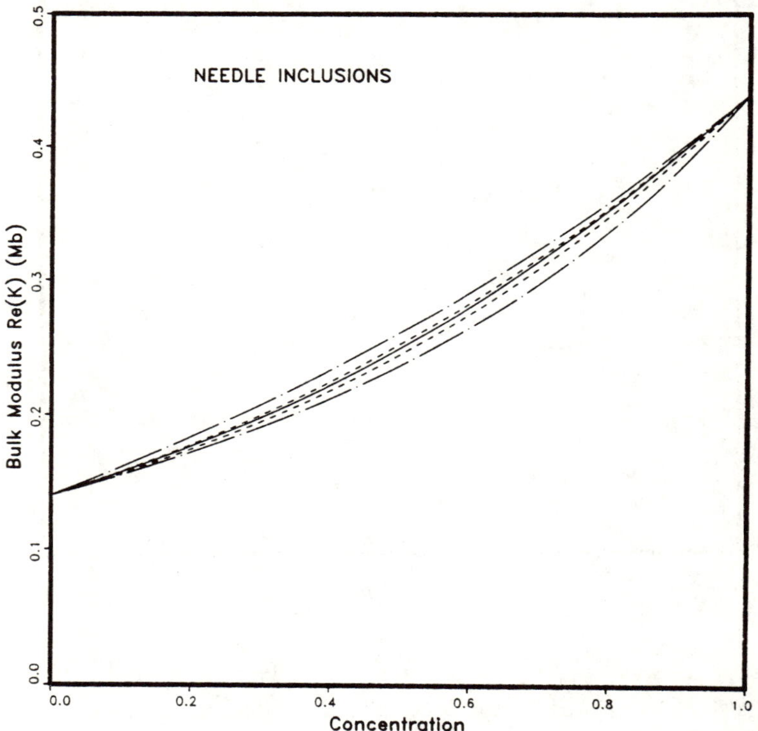


Figure 2(a). Same as Figure 1(a) for needle-shaped inclusions (prolate spheroids with aspect ratio=0).

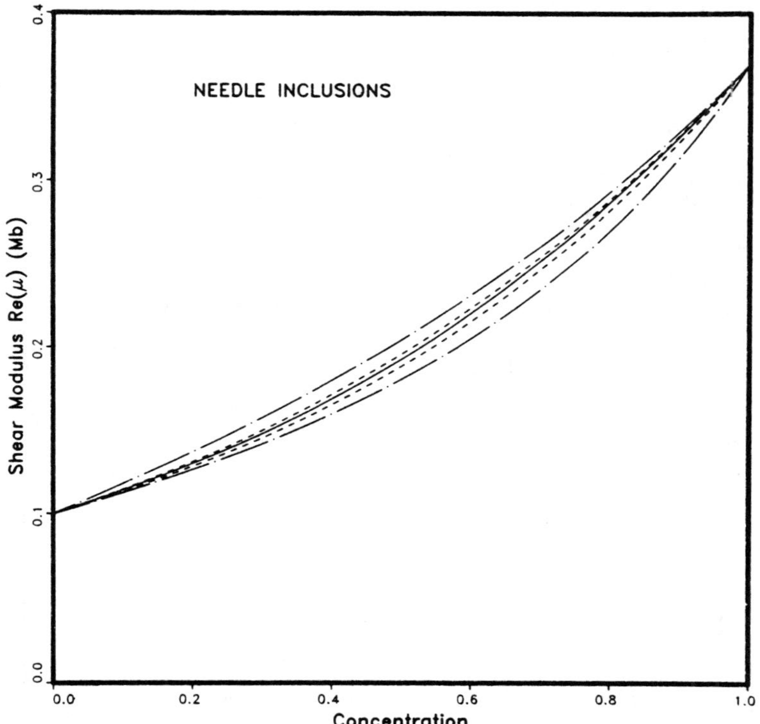


Figure 2(b). Same as Figure 1(b) for needle-shaped inclusions.

Iterating Equation 46, we obtain the Born series

$$\varepsilon = \varepsilon^0 + G\Delta c\varepsilon^0 + G\Delta cG\Delta c\varepsilon^0 + \ldots \tag{48}$$

and then sum the Born series formally to yield

$$\varepsilon = (I+Gt)\varepsilon^0 = (I-G\Delta c)^{-1}\,\varepsilon^0 \tag{49}$$

where the t matrix is defined by

$$t = \Delta c(I-G\Delta c)^{-1} = \Delta c(I+Gt). \tag{50}$$

Taking the ensemble average of Equation 49, we have

$$\langle\varepsilon\rangle = (I+G\langle t\rangle)\varepsilon^{0} = \langle(I-G\Delta c)^{-1}\rangle\varepsilon^{0}. \tag{51}$$

For a single scatterer, Equation 49 is equivalent to Equation 10. Therefore, we note that Wu's tensor T is formally related to the t matrix by

$$T = I+Gt = (I-G\Delta c)^{-1}. \tag{52}$$

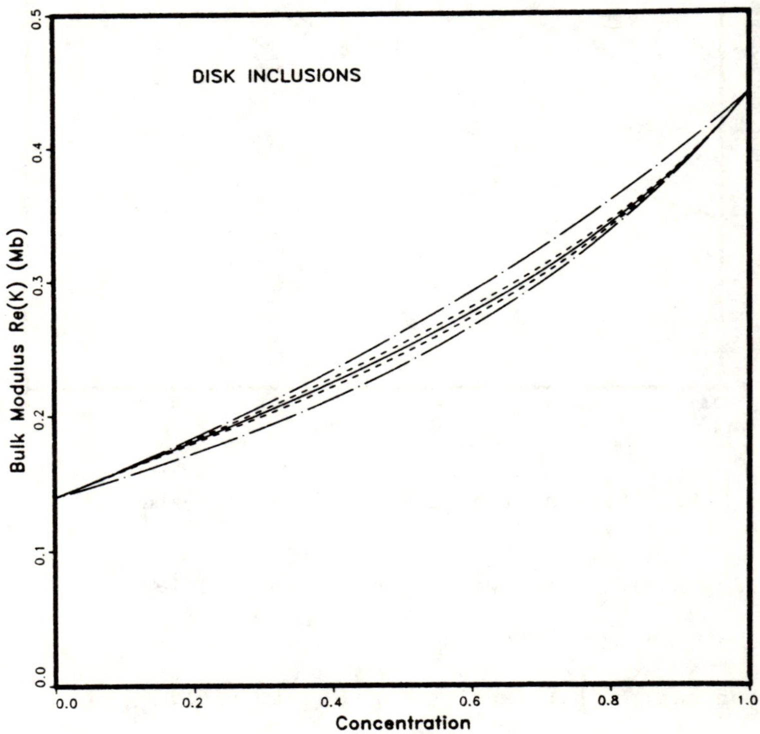


Figure 3(a). Same as Figure 1(a) for disk-shaped inclusions (oblate spheroids with aspect ratio=0).

Equation 51 is now in a convenient form for use in determining the effective elastic tensor c* of a composite defined by

$$\langle\sigma\rangle = \langle c\varepsilon\rangle \equiv c^*\langle\varepsilon\rangle \tag{53}$$

where the averages in Equation 53 are again ensemble averages. Using our standard definition $c = c^m + \Delta c$, we find

$$\begin{aligned}\langle c\varepsilon\rangle &= c^m\langle\varepsilon\rangle + \langle\Delta c\varepsilon\rangle \\ &= c^m\langle\varepsilon\rangle + \langle t\rangle\varepsilon^0.\end{aligned} \tag{54}$$

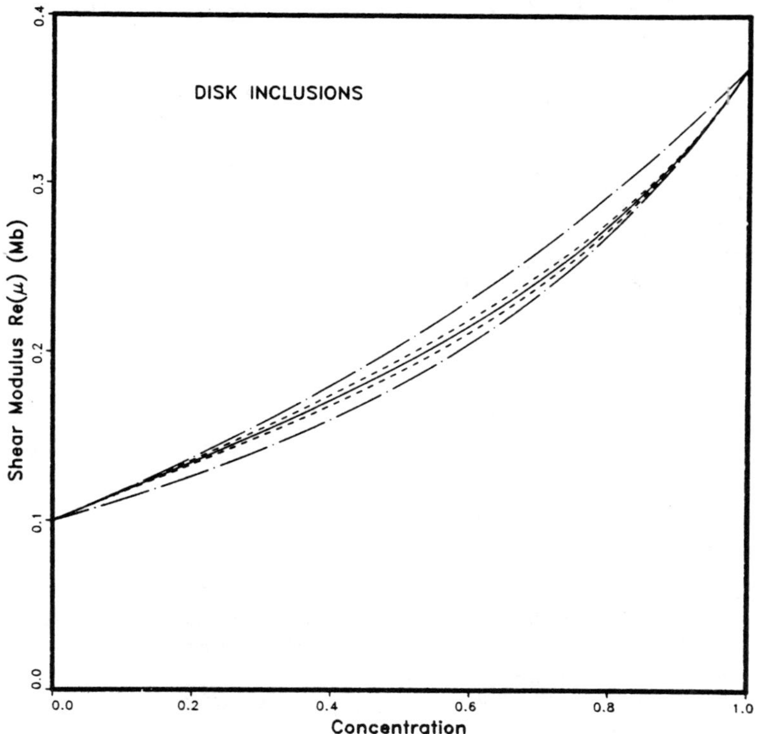


Figure 3(b). Same as Figure 1(b) for disk-shaped inclusions.

From Equation 54, it follows easily that the effective elastic tensor is given by

$$c^* = c^m + \langle t\rangle(I+G\langle t\rangle)^{-1}. \tag{55}$$

The choice of matrix elastic tensor $c^m$ is still completely free since the decomposition $c = c^m + \Delta c$ is not unique. Thus, we are free to choose $c^m = c^*$ which implies

$$\langle t\rangle = 0. \tag{56}$$

Equation 56 is an implicit formula determining the effective elastic tensor $c^*$.

In principle, Equation 56 provides an exact solution for the effective moduli. However, the total t matrix is generally too difficult to calculate. It turns out to be more reasonable and more effective [24] to rearrange the terms of the total t matrix into a series of terms of repeated scattering from individual scatterers ($t_i$). Then, by setting the ensemble average of the individual t matrices to zero

$$\langle t_i\rangle = \sum_{i=1}^{N} f_i\Delta c_i(I-G\Delta c_i)^{-1} = 0 \tag{57}$$

and neglecting terms corresponding to fluctuations in the scattered wave [24] a tractable approximation for the elastic moduli is obtained.

When the constituents and the composite as a whole are all homogeneous and isotropic, the tensor equation 57 reduces to two coupled equations

$$\sum_{i=1}^{N} f_i(K_i - K^*)P^{*i} = 0 \tag{58}$$

$$\sum_{i=1}^{N} f_i(\mu_i - \mu^*)Q^{*i} = 0 \tag{59}$$

where Equations 13, 14 and 52 were used to simplify Equation 57. Note that Equations 58 and 59 are identical to Equations 18 and 19 thereby establishing the equivalence of the two approaches in the isotropic case.

## CONCLUSION

We conclude that our effective medium theory satisfies all the known constraints on a viable theory: (1) it gives correct values and slopes for both large and small volume fractions of inclusions. (2) Numerical evidence indicates that the results always satisfy the Hashin - Shtrikman bounds, the Beran - Molyneux - Miller bounds, and the McCoy - Silnutzer bounds. (3) The theory is known [2] to reproduce Hill's exact result [25] for composites with uniform shear modulus.

The single-scatterer theory is designed to minimize multiple scattering effects while yielding formulas which are relatively easy to use. Nevertheless, the theory is not exact and some potentially significant effects have been neglected. The neglected terms become more important for propagation of higher frequency elastic waves. Future efforts will therefore be directed toward extending the effective medium theory to scattering from clusters of inclusions at finite frequency.

REFERENCES

1. Berryman, J. G. "Theory of Elastic Properties of Composite Materials," *Appl. Phys. Lett.* 35(11):856-858 (1979).

2. Berryman, J. G. "Long-wavelength Propagation in Composite Elastic Media, I. Spherical Inclusions," *J. Acoust. Soc. Am.* 68(6):1809-1819 (1980).

3. Berryman, J. G. "Long-wavelength Propagation in Composite Elastic Media, II. Ellipsoidal Inclusions," *J. Acoust. Soc. Am.* 68(6):1820-1831 (1980).

4. Watt, J. P., Davies, G. F. and O'Connell, R. J. "Elastic Properties of Composite Materials," *Rev. Geophys. Space Phys.* 14(4):541-563 (1976).

5. Elliott, R. J., Krumhansl, J. A. and Leath, P. L. "The Theory and Properties of Randomly Disordered Crystals and Related Physical Systems," *Rev. Mod. Phys.* 46(3):465-543 (1974).

6. Mal, A. K. and Knopoff, L. "Elastic Wave Velocities in Two-Component Systems," *J. Inst. Math. Its Appl.* 3(4):376-387 (1967).

7. Wu, T. T. "The Effect of Inclusion Shape on the Elastic Moduli of a Two-phase Material," *Int. J. Solids Struct.* 2:1-8 (1966).

8. Eshelby, J. D. "The Determination of the Elastic Field of an Ellipsoidal Inclusion, and Related Problems," *Proc. R. Soc. London Ser. A* 241:376-396 (1957).

9. Korringa, J., Brown, R. J. S., Thompson, D. D. and Runge, R. J. "Self-Consistent Imbedding and the Ellipsoidal Model for Porous Rocks," *J. Geophys. Res.* 84(B10):5591-5598 (1979).

10. Hill, R. "The Elastic Behaviour of a Crystalline Aggregate," *Proc. Phys. Soc. London* 65(A):349-354 (1952).

11. Hashin, Z. and Shtrikman, S. "Note on a Variational Approach to the Theory of Composite Elastic Materials," *J. Franklin Inst.* 271(4):336-341 (1961); "On Some Variational Principles in Anisotropic and Nonhomogeneous Elasticity," *J. Mech. Phys. Solids* 10:335-342 (1962); "A Variation Approach to the Theory of the Elastic Behaviour of Multiphase Materials," *J. Mech. Phys. Solids* 11:127-140 (1963).

12. Beran, M. J. and Molyneux, J. "Use of Classical Variational Principles to Determine Bounds for the Effective Bulk Modulus in Heterogeneous Media," *Quart. Appl. Math.* 24:107-118 (1966).

13. McCoy, J. J. In: *Recent Advances in Engineering Science,* A. C. Eringen, Ed. (New York: Gordon and Breach, 1970), pp. 235-254.

14. Miller, M. N. "Bounds for Effective Electrical, Thermal, and Magnetic Properties of Heterogeneous Materials," *J. Math. Phys.* 10(11):1988-2004 (1969), and "Bound for Effective Bulk Modulus of Heterogeneous Materials," *J. Math. Phys.* 10(11):2005-2013 (1969).

15. Silnutzer, N. R. "Effective Constants of Statistically Homogeneous Materials," Ph.D. Thesis, University of Pennsylvania, 283 p., 1972.

16. Milton, G. W. "Bounds on the Electromagnetic, Elastic, and Other Properties of Two-Component Composites," *Phys. Rev. Lett.* 46(8):542-545 (1981).

17. Zeller, R. and Dederichs, P. H. "Elastic Constants of Polycrystals," *Phys. Status Solid.* B55(2):831-842 (1973).

18. Korringa, J. "Theory of Elastic Constants of Heterogeneous Media," *J. Math. Phys.* 14(4):509-513 (1973).

19. Gubernatis, J. E. and Krumhansl, J. A. "Macroscopic Engineering Properties of Polycrystalline Materials: Elastic Properties," *J. Appl. Phys.* 46(5):1875-1883 (1975).

20. Hill, R. "A Self-Consistent Mechanics of Composite Materials," *J. Mech. Phys. Solids* 13:213-222 (1965).

21. Budiansky, B. "On the Elastic Moduli of Some Heterogeneous Materials," *J. Mech. Phys. Solids* 13:223-227 (1965).

22. Walpole, L. J. "On the Overall Elastic Moduli of Composite Materials," *J. Mech. Phys. Solids* 17:235-251 (1969).

23. Boucher, S. "On the Effective Moduli of Isotropic Two-Phase Elastic Composites," *J. Compos. Mater.* 8:82-89 (1974).

24. Velicky, B., Kirkpatrick, S. and Ehrenreich, H. "Single-Site Approximations in the Electronic Theory of Simple Binary Alloys," *Phys. Rev.* 175(3):747-766 (1968).

25. Hill, R. "Elastic Properties of Reinforced Solids: Some Theoretical Principles," *J. Mech. Phys. Solids* 11:357-372 (1963).

CHAPTER 8

# ATTENUATION DUE TO SCATTERING OF ULTRASONIC COMPRESSIONAL WAVES IN GRANULAR MEDIA*

*A. J. Devaney, H. Levine*[+] *and T. Plona*
Schlumberger-Doll Research
P.O. Box 307
Ridgefield, CT 06877, USA

## ABSTRACT

A recently developed theory (A. J. Devaney, *J. Math. Phys.* 21:2603 (1980)) for determining the effective elastic constants of statistically homogeneous random composites is applied to the case of a two-phase composite consisting of a fluid matrix in which are embedded solid (elastic) inclusions (grains). A dispersion relation satisfied by the complex wavenumber $k=k'+ik''$ of coherent compressional waves is obtained for ultrasonic frequencies below that imposed by the Rayleigh limit $k'a < 0.1$ where a is the outer radius of the grains. The dispersion relation is found to depend on the forward scattering amplitude of the grains, the average number of grains per unit volume, and the two-point correlation function for grain centers. The results are applied to a composite of spherical glass beads in a fluid matrix and found to yield values for attenuation ($k''$) that agree with experimental results over wide frequency ranges.

## INTRODUCTION

A theory has been developed recently [1] for determining the wavenumbers $k=k'+ik''$ and $K=K'+iK''$ of the compressional and shear wave coherent modes, respectively, of statistically homogeneous granular composites. At wavenumbers well below the Rayleigh limit ($k'a$ and $K'a<<.1$, with a being the outer grain radius) the imaginary parts of the wavenumbers can be neglected and the theory yields coupled dispersion relations for the real parts of the wavenumbers [2]. These dispersion relations yield, in turn, coupled algebraic equations for the effective bulk and shear moduli of the composite. These later equations were shown in Reference 2 to reduce to those obtained by Kuster and Toksoz [3] in cases where the density of grains is low so that multiple scattering effects can be ignored.

In this paper we shall employ the theoretical model developed in Reference 1 to derive a dispersion relation for the compressional wavenumber k in cases where the effective shear modulus of the composite is negligibly small. Unlike the work reported in Reference 2 we shall consider frequency ranges where the imaginary part k" of the compressional wavenumber is not negligible. Indeed, our main goal is the calculation of the attenuation of compressional waves as a function of frequency for the case of randomly packed spherical grains, fully saturated with a non-viscous fluid.

A brief review of the theoretical model developed in Reference 1 is presented in Section 2. The concept of an effective medium is introduced and is related to the mean elastic wave propagation characteristics of a random composite. Dispersion relations satisfied by the effective compressional and shear wavenumbers of the composite are presented and briefly discussed in this section.

The dispersion relations introduced in Section 2.0 are specialized to the case of composites consisting of spherical grains embedded in a non-viscous fluid in Section 3.0. The compressional wave dispersion relation is found to require the solution of an integral equation that depends on the second order probability distribution for grain center locations. For a completely random [4] distribution of grains the solution to the integral equation is equal to the T matrix [5] for a single scatterer. For this case the dispersion relation reduces to the compressional wave dispersion relation derived in Reference 2 for the case when the effective shear modulus of the composite is negligible.

Section 4 is devoted to a calculation of the compressional wavenumber for cases where correlations between grain center locations are important. The integral equation introduced in Section 3 is solved within the Rayleigh approximation for cases where the correlation length is of the order of the grain diameter. Within this approximation correlations in grain center locations are found to affect only the imaginary part k" of the compressional wavenumber. An expression for k" is derived which depends on the scattering cross section of a single scatterer embedded in the effective medium and on the mean square fluctuation of the total number of grains contained in a unit volume.

The mean square fluctuation in the total number of grains per unit volume is approximately evaluated for a randomly packed distribution of impenetrable spheres. The resulting attenuation factor $\alpha=2k''$ is found to be in excellent agreement with experimentally determined values of attenuation for systems of randomly packed spherical glass beads saturated with water. It is shown that calculations of the attenuation factor based on single scattering theory fail to account for partial coherence of the scattered fields introduced by the intrinsic short range order in densely packed random composites and therefore result in predictions of $\alpha$ far in excess of those observed experimentally.

## DISPERSION RELATIONS FOR THE EFFECTIVE MEDIUM

We consider a composite medium consisting of a random distribution of identical, randomly oriented, elastic inclusions (grains) embedded in a homogeneous, isotropic elastic matrix. In the absence of external body forces the amplitude $u_i(\underline{r})$ of a harmonically oscillating displacement field

$$U_i(\underline{r},t) = u_i(\underline{r})\, e^{-i\omega t} \tag{2.1}$$

propagating in this composite satisfies the wave equation

$$[C_{ijkl}(\underline{r})u_{k,l}(\underline{r})]_{,j} + \omega^2\rho(\underline{r})u_i(\underline{r}) = 0\ . \tag{2.2}$$

In Equation 2.2, $C_{ijkl}(\underline{r})$ is the elastic moduli tensor and $\rho(\underline{r})$ the density at the point $\underline{r}$ in the medium. Because the medium is composed of a random distribution of inclusions both $C_{ijkl}(\underline{r})$ and $\rho(\underline{r})$ are random functions of the position vector $\underline{r}$.

The elastic moduli tensor and density can be expressed in the form

$$C_{ijkl}(\underline{r}) = C^{o}_{ijkl} + \sum_{n=1}^{N} \delta C^{(n)}_{ijkl}(\underline{r})\ , \tag{2.3a}$$

$$\rho(\underline{r}) = \rho^{o} + \sum_{n=1}^{N} \delta\rho^{(n)}(\underline{r}) \tag{2.3b}$$

Here, $C^{o}_{ijkl}$ and $\rho^{o}$ are the elastic moduli tensor and density of the bulk material comprising the matrix and $\delta\, C^{(n)}_{ijkl}$ and $\delta\,\rho^{(n)}$ characterize perturbations in the elastic properties of the composite caused by the presence of an inclusion (grain) centered at the point $\underline{R}_n$. As mentioned above, the grains are assumed to be identical in shape and composition and to be randomly oriented in space. Although N is taken to be finite in Equations 2.3 we shall eventually be interested in the case $N \to \infty$ in such a way that the number of grains per unit volume N/V (V=volume of material) = $\bar{n}$ remains finite.

Now, because the elastic properties of the composite medium vary randomly from point to point, the amplitude $u_i(\underline{r})$ of an elastic wave propagating in this medium will also vary randomly throughout the medium. In many cases, however, the displacement amplitude, although being random, will not differ significantly from its mean value $\langle u_i(\underline{r})\rangle$ as obtained by averaging $u_i(\underline{r})$ over the ensemble of grain locations $\{\underline{R}_n\}$ and grain orientations. For such cases it is then reasonable to ignore the fluctuation in the displacement vector about its mean value and consider the actual random composite to be equivalent to an apparent (effective) medium in which the mean (coherent) wave appears to propagate.

The calculation of the mean value $\langle u_i(\underline{r})\rangle$ of the displacement amplitude and the subsequent identification of an "effective medium" were the principal results obtained in Reference 1. Although the mathematical calculation of $\langle u_i(\underline{r})\rangle$ as presented in Reference 1 is somewhat involved, a very simple physical interpretation of the underlying rationale for the calculation can be given. To see this we consider a sample of the random composite of volume V embedded in an infinite homogeneous, isotropic elastic medium having, as yet, unspecified elastic parameters and density. We shall call this medium surrounding the sample of random composite the "effective medium." Now imagine an elastic plane wave $\psi_i(\underline{r})$ propagating in the effective medium and scattering off the sample of the random composite. The total displacement amplitude (incident plus scattered) in the effective medium is represented symbolically by

$$u_i = \psi_i + G_{ij}T_{jk}\psi_k \ , \qquad (2.4)$$

where $G_{ij}$ is the Green tensor for the effective medium and $T_{jk}$ is the T matrix for the sample of random composite embedded in the effective medium. The second term on the right-hand side of Equation 2.4 represents the field scattered by the sample of random composite. In Reference 1 the elastic parameters and density of the effective medium were chosen so as to cause the average of the scattered field taken over the ensemble of grain locations $\{\underline{R}_n\}$ and grain orientations to vanish. For this choice of effective medium we then have that $\langle u_i(\underline{r})\rangle = \psi_i(\underline{r})$ so that, as far as the mean displacement amplitude is concerned, the sample of random composite is indistinguishable from the effective medium.

The condition that the ensemble averaged scattered field vanish is seen, on using Equation 2.4, to be equivalent to requiring that

$$\langle T_{jk}\rangle \equiv \int d^3R_1 \ldots d^3R_N P(\underline{R}_1 \ldots, \underline{R}_N) E[T_{jk}] = 0. \qquad (2.5)$$

In Equation 2.5, $P(\underline{R}_1 .., \underline{R}_N)$ is the joint probability density function for the grain locations and $E[T_{jk}]$ is the T matrix averaged over all orientations of the grains for the realization $(\underline{R}_1, \ldots \underline{R}_N)$ of the grain centers. The vanishing of $\langle T_{jk}\rangle$

is to be accomplished by appropriate selection of the elastic parameters of the effective medium. Since this medium is assumed to be homogeneous and isotropic there are only three parameters to be determined: the bulk modulus $\kappa^e$, shear modulus $\mu^e$ and density $\rho^e$ [6].

The solution of the equation $<T_{jk}> = 0$ within the quasi-crystalline approximation (QCA) [7] results in the following pair of dispersion relations which must be satisfied by the compressional and shear wavenumbers k and K of the effective medium:

$$(\kappa^o+4/3\mu^o)k^2-\omega^2\rho^o+(2\pi)^3\bar{n}\hat{\underline{a}}_1\cdot\underline{\underline{Q}}(k\hat{\underline{a}}_1,k\hat{\underline{a}}_1)\cdot\hat{\underline{a}}_1 = 0 \ , \qquad (2.6a)$$

$$\mu^o K^2-\omega^2\rho^o+(2\pi)^3\bar{n}\hat{\underline{a}}_2\cdot\underline{\underline{Q}}(K\hat{\underline{a}}_1,K\hat{\underline{a}}_1)\cdot\hat{\underline{a}}_2 = 0 \ . \qquad (2.6b)$$

In these equations $\kappa^o$, $\mu^o$, and $\rho^o$ are the elastic parameters and density of the matrix and $\bar{n} = N/V$ is the number of grains per unit volume (packing density). $\hat{\underline{a}}_1$ and $\hat{\underline{a}}_2$ are orthogonal unit vectors ($\hat{\underline{a}}_i\cdot\hat{\underline{a}}_j = \delta_{ij}$) and the dyadic $\underline{\underline{Q}}(\underline{p},\underline{p}')$ is related through an integral equation to the T matrix for the elastic inhomogeneity $\delta\, C^{(n)}_{ijkl}$, $\delta\,\rho^{(n)}$ centered at the origin $\underline{R}_o=0$ in the effective medium. This integral equation also involves the conditional probability density function $P(\underline{R}/\underline{O})$ for a grain to be located at $\underline{R}$ given that one is located at the origin.

We shall defer until the following section a detailed discussion on the integral equation satisfied by $\underline{\underline{Q}}(\underline{p},\underline{p}')$. For the present we note that in cases where correlations in grain locations can be neglected, $\hat{\underline{a}}_1\cdot Q(k\hat{\underline{a}}_1,k\hat{\underline{a}}_1)\cdot\hat{\underline{a}}_1$ and $\hat{\underline{a}}_2\cdot Q(K\hat{\underline{a}}_1,K\hat{\underline{a}}_1)\cdot\hat{\underline{a}}_2$ are directly related to the compressional and shear wave scattering amplitudes, respectively, produced by the inhomogeneity $\delta\, C^{(n)}_{ijkl}$, $\delta\,\rho^{(n)}$ centered at the origin in the effective medium [1]. For such cases one finds that the dispersion relations (2.6) reduce to a pair of coupled algebraic equations for the elastic parameters $\kappa^e$ and $\mu^e$. The density $\rho^e$ is found to be simply the volume weighted average of the densities of the grains and matrix, a result that follows from simple geometric considerations [2].

## EVALUATION OF THE COMPRESSIONAL WAVE DISPERSION RELATION

In this section we shall focus our attention on a composite consisting of homogeneous spherical grains embedded in a non-viscous fluid. We shall tacitly assume that the effective shear modulus for the composite is negligible ($\mu^e \approx 0$) so that we need only consider the compressional wave dispersion relation (2.6a).

The dyadic $\underline{\underline{Q}}(\underline{p},\underline{p}')$ occurring in the dispersion relations (2.6) was shown in Reference 1 to satisfy the integral equation

$$\underline{Q}(\underline{p},\underline{p}') = \underline{\underline{t}}(\underline{p},\underline{p}') + \int d^3p''\underline{\underline{t}}(\underline{p},\underline{p}'')\cdot\underline{\underline{G}}(\underline{p}'')\cdot\underline{Q}(\underline{p}'',\underline{p}')\tilde{\gamma}(|\underline{p}''-\underline{p}'|) \tag{3.1}$$

In this equation $\underline{\underline{t}}(\underline{p},\underline{p}')$ is the ensemble average, over all scatterer orientations, of the T matrix for the elastic inhomogeneity, $\delta\, C^{(n)}_{ijkl}$, $\delta\, \rho^{(n)}$ centered at the origin in the effective medium. For the present case where the grains are homogeneous spheres $\underline{t}(\underline{p},\underline{p}')$ is simply the T matrix for a homogeneous spherical scatterer. The elastic parameters density of the scatterer are given by [2]

$$\begin{aligned} \kappa &= \kappa' - \kappa^o + \kappa^e \quad , \\ \mu &= \mu' \quad , \\ \rho &= \rho' - \rho^o + \rho^e \quad , \end{aligned} \tag{3.2}$$

where $\kappa'$, $\mu'$ and $\rho'$ are the elastic parameters and density of the grains and $\kappa^o$, $\rho^o$ are those of the non-viscous fluid matrix.

The dyadic $\underline{\underline{G}}(\underline{p}'')$ in Equation 3.1 is the three-fold Fourier transform of the Green dyadic $\hat{\underline{\underline{G}}}(\underline{R})$ for the effective medium. Since $\mu^e \approx 0$, $\underline{\underline{G}}(\underline{R})$ is given by [8]

$$\hat{\underline{\underline{G}}}(\underline{R}) = \frac{1}{4\pi\rho^e\omega^2} \nabla\nabla \frac{e^{ikR}}{R} \tag{3.3}$$

where $\nabla$ is the gradient operator. The Fourier transform of Equation 3.3 is readily obtained and we find that

$$\underline{\underline{G}}(\underline{p}'') = -\frac{p''^2}{\rho^e\omega^2} \left[\frac{\hat{\underline{p}}''\hat{\underline{p}}''}{p''^2 - k^2 - i\varepsilon}\right] \tag{3.4}$$

where $\varepsilon$ is an arbitrarily small positive constant [9] and where $\hat{\underline{p}}''=\underline{p}''/p''$ is the unit vector in the $\underline{p}''$ direction.

Finally, $\tilde{\gamma}(K)$ is given by

$$\tilde{\gamma}(K) = \int d^3R\gamma(R)e^{-i\underline{K}\cdot\underline{R}} \tag{3.5}$$

where

$$\gamma(R) = (N-1)P(\underline{R}|\underline{0}) - \bar{n} \tag{3.6}$$

Here $P(\underline{R}/\underline{0})$ is the conditional probability density function for a grain to be centered at $\underline{R}$ given that one is centered at the origin and $\bar{n}=N/V$ is the mean number of grains per unit volume. Because we have assumed that the grains are spherical and are homogeneously distributed, $\gamma$ depends only on the distance $R = |\underline{R}|$ and, hence, $\tilde{\gamma}$ depends only on the magnitude $K=|\underline{K}|$ of the $\underline{K}$ vector.

On substituting for $\underline{\underline{G}}(\underline{p}'')$ from Equation 3.4 and taking dot products of Equation 3.1 from the left with $\hat{\underline{p}}=\underline{p}/p$ and from the right with $\hat{\underline{p}}'=\underline{p}'/p'$ we obtain

$$Q(\underline{p},\underline{p}')=t(\underline{p},\underline{p}')-\frac{1}{\rho^e\omega^2}\int d^3p''\,\frac{p''^2 t(\underline{p},\underline{p}'')Q(\underline{p}'',\underline{p}')\tilde{\gamma}(|p''-p'|)}{p''^2-k^2-i\varepsilon} \tag{3.7}$$

where

$$Q(\underline{p},\underline{p}') = \hat{\underline{p}}\cdot\underline{\underline{Q}}(p,p')\cdot\hat{p}' \quad , \tag{3.8a}$$

$$t(\underline{p},\underline{p}') = \hat{\underline{p}}\cdot\underline{\underline{t}}(\underline{p},\underline{p}')\cdot\hat{\underline{p}}' \quad . \tag{3.8b}$$

For future reference we note that when $p=p'=k$, the quantity $t(\underline{p},\underline{p}')$ is directly related to the compressional wave scattering amplitude produced by the spherical inclusion whose elastic parameters and density are given in Equations 3.2. In particular, as $k'r \to \infty$ the amplitude $\underline{u}^{(s)}(\underline{r})$ of the compressional wave scattered by this inclusion will be of the form

$$\underline{u}^{(s)}(\underline{r})\sim\hat{\underline{r}}f(\hat{\underline{r}},k\underline{s}_o)\,\frac{e^{ikr}}{r} \quad , \tag{3.9}$$

where $k\underline{s}_o$ is the wavenumber of the incident plane wave to the inclusion. The scattering amplitude $f(\hat{\underline{r}},k\underline{s}_o)$ is then related to $t(k\hat{\underline{r}},k\underline{s}_o)$ via the equation [1]

$$t(k\hat{\underline{r}},k\underline{s}_o) = -\frac{\rho^e\omega^2}{2\pi^2k^2}f(\hat{\underline{r}},k\underline{s}_o). \tag{3.10}$$

The scattering amplitude $f(\hat{\underline{r}},k\underline{s}_o)$ can be evaluated for homogeneous spherical inclusions and one finds that within the Rayleigh approximation [10]

$$f(\hat{\underline{r}},k\underline{s}_o) = \frac{k^2v}{4\pi}\left(\frac{\delta\rho}{\rho^e}\cos\Theta - \frac{\delta\kappa}{\kappa^e+\delta\kappa}\right) \quad , \tag{3.11}$$

so that

$$t(k\hat{\underline{r}},k\underline{s}_o) = -\frac{\rho^e\omega^2 v}{(2\pi)^3}\left(\frac{\delta\rho}{\rho^e}\cos\theta - \frac{\delta\kappa}{\kappa^e+\delta\kappa}\right). \qquad (3.12)$$

In the above equations $v = 4/3\pi a^3$ is the volume of the scatterer ($a$ = grain radius), $\delta\kappa = \kappa'-\kappa^o$, $\delta\rho = \rho'-\rho^o$ and $\theta$ is the angle between the unit vectors $\hat{\underline{r}}$ and $\underline{s}_o$.

The dispersion relation (2.6a) for the compressional wavenumber $k$ can be expressed in terms of $Q(k\hat{\underline{a}}_1,k\hat{\underline{a}}_1)$ as follows:

$$\kappa^o k^2-\omega^2\rho^o+(2\pi)^3\bar{n}Q(k\hat{\underline{a}}_1,k\hat{\underline{a}}_1) = 0. \qquad (3.13)$$

Assuming that the imaginary part $k''$ of the wavenumber $k$ is small in comparison with the real part $k'$, we find from Equation 3.13 that $k'$ and $k''$ satisfy the following two equations [11]:

$$\kappa^o k'^2-\omega^2\rho^o+(2\pi)^3\bar{n}Q'(k'\hat{\underline{a}}_1,k'\hat{\underline{a}}_1) = 0, \qquad (3.14a)$$

$$k'' = -\frac{(2\pi)^3\bar{n}}{2\kappa^o k'}Q''(k'\hat{\underline{a}}_1,k'\hat{\underline{a}}_1). \qquad (3.14b)$$

In Equations 3.14 $Q'$ and $Q''$ are the real and imaginary parts, respectively, of $Q$. Equation 3.14a can be regarded as a dispersion relation for $k'$ while Equation 3.14b is an expression for the imaginary part of the wavenumber in terms of its real part.

The lowest order solution of Equation 3.7 is obtained by simply neglecting the second term on the right-hand side of the equation so that $Q(k'\hat{\underline{a}}_1,k'\hat{\underline{a}}_1)\approx t(k'\hat{\underline{a}}_1,k'\hat{\underline{a}}_1)$. This was precisely the approximation employed in Reference 2 to calculate the effective elastic parameters and density of granular composites at frequencies well below the Rayleigh limit $k'a=0.1$. Within this approximation Equations 3.14 become

$$\kappa^o k'^2-\omega^2\rho^o+(2\pi)^3\bar{n}t'(k'\hat{\underline{a}}_1,k'\hat{\underline{a}}_1) = 0, \qquad (3.15a)$$

$$k'' = -\frac{(2\pi)^3\bar{n}}{2\kappa^o k'}t''(k'\hat{\underline{a}}_1,k'\hat{\underline{a}}_1). \qquad (3.15b)$$

The expressions for the effective bulk modulus and density given in Reference 2 are readily obtained from Equation 3.15a upon substituting for $t(k'\hat{\underline{a}}_1,k'\hat{\underline{a}}_1)$ from Equation 3.12. On making this substitution we obtain the result

$$[\kappa^o + C \frac{\kappa^e \delta\kappa}{\kappa^e + \delta\kappa}]\, k'^2 - \omega^2 [\rho^o + C\delta\rho] = 0, \tag{3.16}$$

where $C=\bar{n}v$ is the volume concentration of grains. On comparing Equation 3.16 with the standard form of the dispersion relation for non-absorbing homogeneous, isotropic media

$$\kappa^e k'^2 - \omega^2 \rho^e = 0 , \tag{3.17}$$

we conclude that

$$\kappa^e = \kappa^o + C \frac{\kappa^e \delta\kappa}{\kappa^e + \delta\kappa} , \tag{3.18a}$$

$$\rho^e = \rho^o + C\, \delta\rho . \tag{3.18b}$$

These equations are precisely the results obtained in Reference 2 for the limiting case when $\mu^e \approx 0$.

We shall show below that Equation 3.15a continues to hold within the next higher order approximation to $Q(k\hat{\underline{a}}_1, k\hat{\underline{a}}_1)$ but that Equation 3.15b needs to be modified. The modifications in Equation 3.15b is necessitated by the intrinsic short range order found in densely packed granular composites. This intrinsic order has the net effect of decreasing the attenuation due to scattering and hence, results in values of k" smaller than that predicted by Equation 3.15b.

The next level of approximation to $Q(\underline{p},\underline{p}')$ is obtained by substituting $t(\underline{p}'',\underline{p}')$ for $Q(\underline{p}'',\underline{p}')$ in the integral in Equation 3.7. Performing this substitution and setting $\underline{p}=\underline{p}'=k\hat{\underline{a}}_1$ we obtain

$$Q(k\hat{\underline{a}}_1, k\hat{\underline{a}}_1) \approx t(k\hat{\underline{a}}_1, k\hat{\underline{a}}_1)$$
$$- \frac{1}{\rho^e \omega^2} \int d^2p'' \, \frac{p''^2 t(k\hat{\underline{a}}_1, \underline{p}'') t(\underline{p}'', k\hat{\underline{a}}_1) \tilde{\gamma}(|\underline{p}'' - k\hat{\underline{a}}_1|)}{p''^2 - k^2 - i\varepsilon} \tag{3.19}$$

The function $(p''^2 - k^2 - i\varepsilon)^{-1}$ can be decomposed into a principal valued part and a delta function contribution as follows [12]:

$$\frac{1}{p''^2 - k^2 - i\varepsilon} \equiv \frac{P}{p''^2 - k^2} + i\pi\delta(p''^2 - k^2) . \tag{3.20}$$

On substituting Equation 3.20 into 3.19 we obtain

$$Q(k\hat{\underline{a}}_1, k\hat{\underline{a}}_1) \approx t(k\hat{\underline{a}}_1, k\hat{\underline{a}}_1) + F(k)$$

$$- \frac{i\pi k}{2\kappa^e} \int d\Omega_s t(k\hat{\underline{a}}_1, k\underline{s}) t(k\underline{s}, k\hat{\underline{a}}_1) \tilde{\gamma}(k|\underline{s}-\hat{\underline{a}}_1|), \tag{3.21}$$

where we have used Equation 3.17 and where F(k) is the principal valued integral

$$F(k) = -\frac{1}{\rho^e \omega^2} P \int d^3p'' \frac{p''^2 t(k\hat{\underline{a}}_1, \underline{p}'') t(\underline{p}'', k\hat{\underline{a}}_1) \tilde{\gamma} |p'' - k\hat{\underline{a}}_1|}{p''^2 - k^2} \tag{3.22}$$

Now, the integral F(k) depends only on the T matrix $t(\underline{p},\underline{p}')$ for values of $\underline{p}$ and $\underline{p}'$ such that $p \neq p'$. These "off-the-shell" components of the T matrix can be expected to go to zero rapidly with increasing $|\underline{p}-\underline{p}'|$ so that F(k) can be expected to be significantly smaller than the two terms on the right-hand side of Equation 3.21 [13]. We accordingly shall neglect the contribution of F(k) to $Q(k\hat{\underline{a}}_1, k\hat{\underline{a}}_1)$ in this equation. On replacing k by k' and taking the real and imaginary parts of the resulting expression we then find that

$$Q'(k'\hat{\underline{a}}_1, k'\hat{\underline{a}}_1) \approx t'(k'\hat{\underline{a}}_1, k'\hat{\underline{a}}_1), \tag{3.23a}$$

$$Q''(k'\hat{\underline{a}}_1, k'\hat{\underline{a}}_1) \approx t''(k'\hat{\underline{a}}_1, k'\hat{\underline{a}}_1)$$

$$- \frac{\pi k'}{2\kappa^e} \int d\Omega_s |t(k'\underline{s}, k'\hat{\underline{a}}_1)|^2 \tilde{\gamma}(k'|\underline{s}-\hat{\underline{a}}_1|) \tag{3.23b}$$

In deriving the above expressions we have made use of the facts that $t(k'\hat{\underline{a}}_1, k'\underline{s}) = t^*(k'\underline{s}, k'\hat{\underline{a}}_1)$ and that $\tilde{\gamma}(K)$ is real.

We see from Equation 3.23a that, as discussed earlier, the real part of Q remains unchanged within this next level of approximation. It follows that Equation 3.15a and, hence, Equations 3.18 for the effective bulk modulus and density continue to hold within this approximation. Equation 3.15b is no longer valid, however, and must be replaced by Equation 3.14b with Q" given by (3.23b). We shall evaluate k" within this approximation in the following section.

## IMAGINARY PART OF THE COMPRESSIONAL WAVENUMBER

To evaluate Q" as defined in Equation 3.23b we first note that in a randomly packed composite $\gamma(R)$ will be a rapidly decreasing function for values of R greater than the correlation length for the composite. The correlation length will be on the order of the grain diameter 2a so that for wavenumbers below the Rayleigh limit k'a=0.1 the quantity $\tilde{\gamma}[k'|\underline{s}-\hat{\underline{a}}_1|]$ is approximately equal to

$$\tilde{\gamma}[k'|\underline{s}-\underline{a}_1|] \approx \tilde{\gamma}_o \equiv \int_V d^3R\gamma(R). \qquad (4.1)$$

On substituting Equation 4.1 into Equation 3.23b we obtain the following approximation to Q":

$$Q''(k'\hat{\underline{a}}_1,k'\hat{\underline{a}}_1) \approx t''(k'\hat{\underline{a}}_1,k'\hat{\underline{a}}_1)$$

$$-\frac{\pi k'}{2\kappa^e}\tilde{\gamma}_o \int d\Omega_s |t(k'\underline{s},k'\hat{\underline{a}}_1)|^2 . \qquad (4.2)$$

Substituting for $t(k'\underline{s},k'\hat{\underline{a}}_1)$ from Equation 3.10 we find that

$$\int d\Omega_s |t(k'\underline{s},k'\hat{\underline{a}}_1)|^2 = \left(\frac{\rho^e\omega^2}{2\pi^2k'^2}\right)^2 \int d\Omega_s |f(\underline{s},k'\hat{\underline{a}}_1)|^2. \qquad (4.3)$$

The integral on the right-hand side of Equation 4.3 can be shown to be equal to the total scattering cross section $\sigma^s$ associated with the scattering amplitude $f(\underline{s},k'\hat{\underline{a}}_1)$ [14]. Moreover, it follows from the optical theorem [14] that this integral is directly proportional to the imaginary part of the forward scattering amplitude $f(\hat{\underline{a}}_1,k'\hat{\underline{a}}_1)$;

$$\int d\Omega_s |f(\underline{s},k'\hat{\underline{a}}_1)|^2 = \sigma^s = \frac{4\pi}{k'} \operatorname{Im} f(\hat{\underline{a}}_1,k'\hat{\underline{a}}_1). \qquad (4.4)$$

Finally, on substituting for $t''(k'\hat{\underline{a}}_1,k'\hat{\underline{a}}_1)$ from Equation 3.10 and making use of Equation 4.4 we find that (4.2) reduces to

$$Q''(k'\hat{\underline{a}}_1,k'\hat{\underline{a}}_1) \approx -\frac{\rho^e\omega^2}{2\pi^2k'^2}(1+\tilde{\gamma}_o)\operatorname{Im} f(\hat{\underline{a}}_1,k'\hat{\underline{a}}_1). \qquad (4.5)$$

The imaginary part of the wavenumber is obtained by substituting the above expression for Q" into Equation 3.14b. We find that

$$k'' = \frac{2\pi\bar{n}}{k'}(1+\tilde{\gamma}_o)\operatorname{Im} f(\hat{\underline{a}}_1,k'\hat{\underline{a}}_1)$$

$$= \frac{\bar{n}}{2}(1+\tilde{\gamma}_o)\sigma^s, \qquad (4.6)$$

where we have used Equation 4.4. We can express k" in terms of $\kappa^e$, $\rho^e$, $\delta\kappa$ and $\delta\rho$ by evaluating Equation 4.4 for $\sigma^s$ using the expression for the scattering amplitude $f(\hat{\underline{r}},k\underline{s}_o)$ given in

Equation 3.11. We find that

$$\sigma^s = \frac{k'^4 v^2}{2\pi}\left[\frac{1}{3}\left(\frac{\delta\rho}{\rho^e}\right)^2 + \left(\frac{\delta\kappa}{\kappa^e+\delta\kappa}\right)^2\right] \tag{4.7}$$

so that

$$k'' = \frac{\bar{n}(1+\tilde{\gamma}_o)k'^4 v^2}{4\pi}\left[\frac{1}{3}\left(\frac{\delta\rho}{\rho^e}\right)^2 + \left(\frac{\delta\kappa}{\kappa^e+\delta\kappa}\right)^2\right] . \tag{4.8}$$

In the lowest order approximation k" is given by Equation 3.15b which, on making use of Equation 3.10, reduces to

$$k'' \approx \frac{2\pi n}{k'} \operatorname{Im} f(\hat{\underline{a}}_1, k'\hat{\underline{a}}_1) = \frac{\bar{n}}{2}\sigma^s . \tag{4.9}$$

The two expressions (4.6) and (4.9) are seen to differ only by the factor $1+\tilde{\gamma}_o$. Since Equation 4.9 holds in the limiting case of $\bar{n}=0$ we must require that $\tilde{\gamma}_o=0$ in this limit.

The expression for k" given in Equation 4.9 for the limiting case of $\tilde{\gamma}_o=0$ has a simple physical interpretation. To see this we note that the total power loss P experienced by a unit amplitude plane wave propagating through a volume V of homogeneous material having a wavenumber $k=k'+ik''$ is equal to $2Vk''$. We conclude then on using Equation 4.9 that for small grain concentrations

$$P = 2Vk'' \approx N\sigma^s \tag{4.10}$$

where $N=\bar{n}V$ is the total number of grains in V. Equation 4.10 states that the total power loss to scattering experienced by a plane wave propagating through a volume V of a granular composite is simply the total number of grains in V times the power loss due to scattering from a single grain [15]. This result implies that the grains scatter <u>incoherently</u>. For large values of n the scattering will no longer be incoherent due to the correlation in grain locations that exists at high grain concentrations. This correlation manifests itself in a negative value of $\tilde{\gamma}_o$ which reduces k" and, hence, the total power loss experienced by acoustic waves propagating in the composite.

To determine k" for large values of $\bar{n}$ we must evaluate the integral of $\gamma(R)$ over the volume V. Unfortunately, there appears to be no closed form expression available in the literature for the conditional density function $P(\underline{R}/\underline{O})$ for random distributions of closely packed, impenetrable spheres. Attempts at numerically integrating $\gamma(R)$ using numerically determined expressions for $P(\underline{R}/\underline{O})$ have not been successful due to the oscillatory behavior of this function for values of $R > 2a$. The values of $\gamma(R)$ for large values of R contribute to the integral (4.1) due to the presence of the factor $R^2$ in the differential volume element $d^3R$.

An alternative method for evaluating $\tilde{\gamma}_o$ follows from the observation that [16]

$$\tilde{\gamma}_o = \frac{1}{<N>} [<N^2> - <N>^2] \qquad (4.11)$$

where N is the total number of grains in the volume V. On substituting Equation 4.11 into (4.6) we obtain the result

$$\begin{aligned} k'' &= \frac{\bar{n}}{2} \frac{1}{<N>}[<N^2> - <N>^2]\, \sigma^s \\ &= [\frac{<N^2>-<N>^2}{V}] \frac{\sigma^s}{2} , \end{aligned} \qquad (4.12)$$

where we have used the fact that $\bar{n}=<N>/V$.

The quantity $(<N^2>-<N>^2)/V$ is the mean square fluctuation in the number of grains per unit volume. This quantity cannot be calculated exactly for a random distribution of densely packed impenetrable spheres. However, we shall approximate the system of randomly packed spheres by a close packed system having missing spheres [17]. Then a simple calculation gives the result that

$$\frac{<N^2>-<N>^2}{V} = \bar{n}(1-p) \quad , \qquad (4.13)$$

where p is the probability of a given lattice site in the close packed system being occupied. Substituting Equation 4.13 into (4.12) then yields the following expression for k":

$$k'' = \frac{\bar{n}}{2} (1-p)\sigma^s \ . \qquad (4.14)$$

We note that for a perfect lattice $p=1$ and $k''=0$. This result is reasonable because at wavenumbers below the Rayleigh limit the only Bragg diffraction peak for a perfect lattice occurs in the forward direction so that all the energy is coherently scattered and there is no attenuation due to scattering. In the limit of a completely random system $p=0$, and we recover the expression (4.9) for this case. For intermediate values of p we shall take p to be the ratio of the volume concentration of grains $C=\bar{n}V$ to the volume concentration of spheres occurring in a perfect close packed lattice. This choice for p insures that the average number of grains in the system is correct. Since the volume concentration for a perfect close packed system is 0.74 we conclude that

$$k'' = \frac{\bar{n}}{2} (1 - \frac{C}{.74}) \sigma^s . \tag{4.15}$$

We have applied the model given in Equation 4.15 to calculate the attenuation of compressional plane waves in randomly packed spherical glass beads saturated by water. For this sytem the volume concentration C was 0.62. On using the expression for $\sigma^s$ given in Equation 4.7 we find that Equation 4.15 yields the following expression for the attenuation constant $\alpha = 2k''$:

$$\alpha = 0.016\ k'^4 v \left[ \frac{1}{3} \left( \frac{\delta\rho}{\rho^e} \right)^2 + \left( \frac{\delta\kappa}{\kappa^e + \delta\kappa} \right)^2 \right]. \tag{4.16}$$

In this expression $\kappa^e$ and $\rho^e$ are given by Equations 3.18 and k' is obtained from Equation 3.17 using the calculated values for $\kappa^e$ and $\rho^e$.

Plots of the attenuation factor $\alpha$ in DB/inch as calculated from Equation 4.16 are shown in Figure 1 for three different grain radii. The experimentally determined values of attenuation are seen to agree extremely well with the theoretical model throughout the frequency ranges shown. It should be noted that simpler models for attenuation which either are not self consistent or which ignore correlations (e.g., Equation 4.9) predict values for $\alpha$ four to five times greater than those observed experimentally.

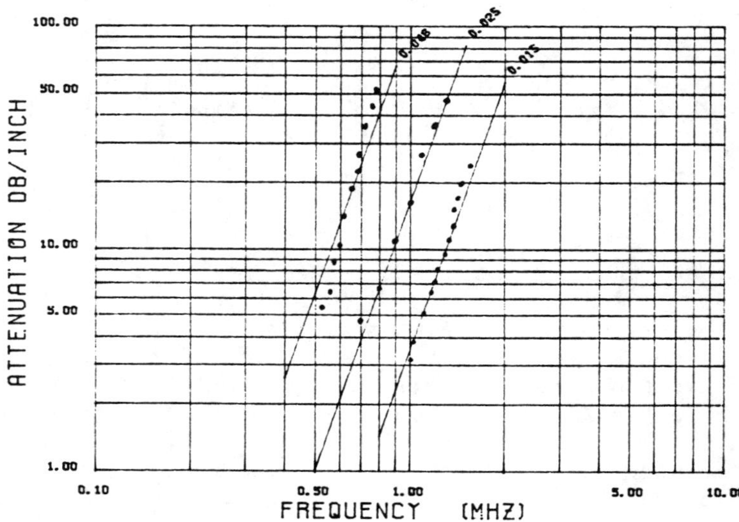


Figure 1. Attenuation factor $\alpha$ as a function of frequency for glass spheres of radii .046, .025 and .015 millimeter.

*A preliminary account of some of the results reported here were presented at the 99th meeting of the Acoustical Society of America (*J. Acoust. Soc. Am.* 67:S44 (1980)).

†Currently with the Dept. of Physics, Harvard Univ., Cambridge, Mass. 02138.

REFERENCES

1. Devaney, A. J. "Multiple scattering theory for discrete, elastic, random media," *J. Math. Phys.* 21(11):2603-2611 (1980).

2. Devaney, A. J. and Levine, H. "Effective elastic parameters of random composites," *Appl. Physics Lett.* 37(4): 377-379 (1980).

3. Kuster, G. T. and Toksoz, M. H. "Velocity and attenuation of seismic waves in two phase media Part I, Theoretical formulations," *Geophysics* 39(5):587-606 (1974); "Part II, Experimental Results," 39(5):607-618 (1978).

4. A "completely random" distribution of grains is one for which the grain locations are statistically independently distributed. Although the locations of a system of impenetrable grains are correlated, the assumption of statistical independence is a good approximation for low volume concentrations of grains.

5. For definition of *T Matrix* (Transition Matrix) see Newton, R. G., *Scattering Theory of Waves and Particles* (New York: McGraw-Hill, 1966), p. 185.
The T matrix is discussed in connection with elastic wave scattering in Reference 1 and in Varadan, V. K., Varadan, V. V., and Yin-Hsing Pao, "Multiple scattering of elastic waves by cylinders of arbitrary cross sections - I. S-H waves," *J. Acoust. Soc. Am.* 63(5):1310-1319 (1978). Its use in acoustic applications is discussed by Waterman, P. C., "New formulation of Acoustic Scattering," *J. Acoust. Soc. Am.* 45(6):1417-1429 (1969).

6. Although the effective medium is homogeneous and isotropic it can be spatially dispersive. This means that it may possess more than one compressional or shear wave number and, hence, more than one bulk and elastic modulus. For further discussions of spatial dispersion in elastic composites see References 1 and 2 and Devaney, A. J. "Spatial dispersion in porous elastic media," *J. Acoust. Soc. Am.* 68:S-15 (1980).

7. The quasicrystalline approximation was first introduced by Lax, M. "Multiple scattering of waves," *Rev. Mod. Phys.* 23(4):287-310 (1951). The self consistent formulation of the QCA employed in Reference 1 is due to Gyorffy, B. L. "Electronic states in liquid metals: A Generalization of Coherent-Potential Approximation for System with Short Range Order," *Phys. Rev. B1* 1(8):3290-3299 (1970) and Korringa, J. and Mills, R. L. "Coherent-Potential Approximation for Random System with Short-Range Correlations," *Phys. Rev. B* 5(4):1654-1656 (1972).

8. Gubernatis, J. E., Domany, E. and Krumhansl, J. A. "Formal aspects of theory of the scattering of ultrasound by flaws in elastic materials," *J. Appl. Phys.* 48(7):2804-2811 (1977). The expression for the green dyadic given in this reference differs in sign from the expression given in Equation 3.3. This is due to the trivial difference in the definitions of green dyadic employed in this reference and in Reference 1.

9. The presence of the positive constant $\varepsilon$ in Equation 3.4 guarantees that the green dyadic obeys the Sommerfeld radiation condition. See, for example the discussion in Jackson, J. D., *Classical Electrodynamics* (New York: John Wiley, 1962) See 6.6.

10. Gubernatis, J. E., Krumhansl, J. A. and Thomson, R. M. "Interpretation of elastic wave scattering theory for analysis and design of flow characterization experiments. The long wave length limit," *J. Appl. Phys.* 50(5):3338-3345 (1979).

11. The assumption that $k'' \ll k'$ is not necessary for the calculation of the wave number from Equation 3.13. In the cases considered in this paper this condition is, however, satisfied and since it simplifies the analysis it will be employed throughout the remainder of the paper.

12. See, for example, Newton, R. G., *Scattering Theory of Waves and Particles* (New York: McGraw-Hill, 1966), p. 178.

13. A more rigorous justification for this approximation can be obtained by direct contour integration over one of the Cartesian components of $\underline{p}''$ in the integral appearing in Equation 3.19. One finds that in the long wave length limit this integral is given approximately by the integral appearing in Equation 3.21.

14. Gubernatis, J. E., Domany, E. and Krumhansl, J. A. "Formal aspects of the theory of the scattering of ultrasound by flaws in elastic materials," *J. Appl. Phys.* 48 48(7):2804-2811 (1977).

15. Strictly speaking $\sigma^s$ is the scattering cross section for an inclusion having the elastic parameters and density given in Equation 3.2 and is <u>not</u> the cross section for a grain of the random composite. However, for small values of $\bar{n}$ $\kappa^o \approx \kappa^e$, and $\rho^o \approx \rho^e$ so that $\kappa \approx \kappa'$ and $\rho \approx \rho'$ and $\sigma^s$ does reduce to the scattering cross section for a grain of the composite.

16. Ziman, J. M., *Models of Disorder* (New York: Cambridge Univ. Press, 1979) p. 178.

17. By a close packed system we mean a rectangular lattice of spheres arranged so as to yield the maximum volume concentration possible in a rectangular lattice structure. See Reference 16, Chap. 1.

CHAPTER 9

# CLASSIFICATION OF SUBMERGED TARGETS BY ACOUSTIC MEANS

*S. K. Numrich, Louis R. Dragonette and Lawrence Flax*
Naval Research Laboratory,
Washington, D.C. 20375

## ABSTRACT

The basis for the classification of known submerged targets has been established through previous analytic and empirical work. The form function (backscattered pressure vs frequency) is unique for a body of a given shape and composition, and correlation with an analytically or empirically obtained form function signature can be accomplished in real time with present computer capabilities.

The relationship between the free body resonances of a body and its signature have been determined empirically and formalized mathematically for bodies of simple shape. An incident acoustic wave couples strongly to many of these resonances and their frequency position and amplitude can be obtained with great precision. Material constants have been obtained to within 1% from scattering measurements made with solid spherical and cylindrical targets, and analytic predictions indicate that at least an order of magnitude better resolution can be achieved.

The resonance theory is particularly useful in the prediction of elastic behavior at the low end of the frequency spectrum. For smooth finite targets the Rayleigh surface wave resonance is the lowest frequency, strongly coupled elastic scattering effect observed. The form function of an elastic body is accurately simulated by the form function of a body with no elasticity at frequencies below the Rayleigh resonance. Predictions of the frequency position of the lowest frequency resonance can be made through simple geometric considerations, and comparisons between resonance positions predicted in this fashion and T-matrix computations for an elastic prolate spheroid show excellent agreement.

The most significant contribution of the resonance formalism; however, has as yet not been exploited. The separation of the scattering by a submerged target into resonance and geometric contributions is not simply a mathematical artifact but something which can be achieved experimentally, and thus be a tool in the classification of unknown targets. Initial experiments have been completed on the reflection from finite, air filled cylindrical shells. Data processing techniques have been used to extract the geometric beam pattern of the target from the far more complicated elastic patterns, and comparisons of these preliminary results with the geometric beam patterns predicted by T-matrix theory are promising.

## THEORETICAL AND EXPERIMENTAL BACKGROUND

The ability to identify and localize submerged bodies by acoustic means became an absolute necessity with the advent of the submarine. Military interest in sonar capability continued as the submarine became a major factor in naval warfare. In recent years, however, interest in acoustic interrogation of bodies contained in a fluid medium has spread to such areas as medical diagnosis and non-destructive evaluation. The primary problem in analyzing sound waves as they interact with targets in fluids is the presence of elastic properties of the materials involved. Developments in computer technology, particularly in the areas of digital signal processing and matrix operations, have pushed research in acoustic scattering to levels of greater competence in understanding and modeling material properties. Real time localization and identification of submerged targets is beginning to appear as a realizable goal.

The major challenge in analytical modeling is to formulate the acoustic scattering problem in such a way as to permit numerical evaluation. Some of the earliest attempts to meet this challenge did so by using as targets simple shapes whose symmetry or infinite extent reduced the computations to a manageable form.

Exact normal mode series solutions to describe acoustic scattering by elastic spheres and infinite cylinders were developed by Faran [1,2] in 1951. Though limited by the then existing computational machines, he evaluated the results for $ka < 5$ and predicted the strong association between elastic scattering and the free modes of vibration of the targets. In 1961, Hickling [3] brought the digital computer to bear significantly on the scattering problem. He extended Faran's calculations to ka values of 30 and computed the normalized, steady-state, backscattered pressure as a function of ka for a series of spherical targets immersed in water [3,4]. These computed curves of acoustic pressure vs frequency constitute the form function of the target body. Since the form function is unique for each material and shape, it is the essence of a classification tool.

Form function curves are shown in Figure 1 for a series of solid spheres. The first curve is the response of a rigid sphere. The form function quickly approaches the value of one where it remains with only minor oscillations. Thus the form function of the rigid sphere at high values of ka is effectively one. The remaining curves illustrate the great variations caused by different material properties. They are arranged in increasing degrees of elasticity beginning with tungsten carbide, the most rigid of the materials, followed by aluminum and lucite. The differences between the form functions of the two metals are clearly discernible and the differences between the metals and the plastic are dramatic.

Before the form function could be given serious consideration as a practical classification tool, it was necessary to demonstrate that this function could be obtained experimentally and that it could be measured accurately in a real time framework. These problems were addressed in three papers [5-7] published at the Naval Research Laboratory (NRL). Neubauer et al. [5] demonstrated that the computed form function for solid, elastic spheres was in excellent agreement with the form function obtained from steady-state, single frequency measurements. Dragonette et al. [6] showed that a significant portion of the form function curve could be obtained accurately from one broadband measurement, and Dardy et al. [7] further indicated that such a broadband measurement could be taken and analyzed in real time.

Identification or rejection of a submerged body as a target of known size, shape and material can be accomplished in real time by the correlation of the measured form function of the submerged body with the form function for the sought target. This can only be done if the form function for the target of interest had been determined either empirically or analytically and is available to the computer. Figure 2 illustrates this type of comparison. The analytically computed form function for a solid tungsten carbide sphere, shown as the solid curve, is used as the basis of comparison with the measured form function displayed as points. The measurements were made using the digital data acquisition and analysis system at NRL. The entire ka range, from 14 to 70, was covered in two experiments performed with short, broadband incident pulses.

At the Catholic University of America, an analytic program generally referred to as the creeping wave theory was developed by Uberall and collaborators [8-10]. The normal mode series solutions for elastic scattering in an infinite cylindrical geometry was transformed into an integral formulation using the Sommerfeld-Watson transformation [11]. One of the original reasons for attempting the creeping wave method was to allow scattering computation in the region of high ka, thus avoiding difficulties inherent in dealing with

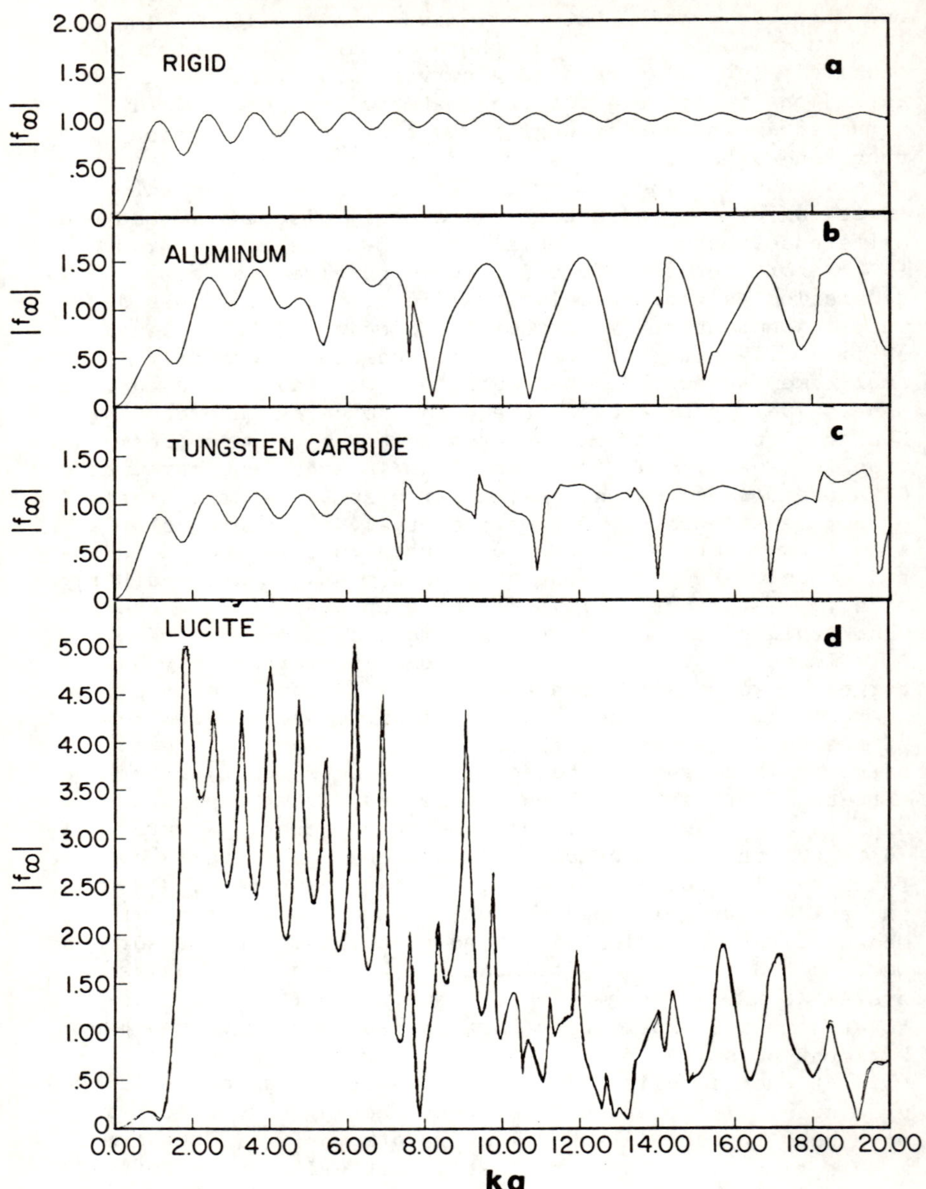


Figure 1. Form functions for (a) a rigid sphere (top), (b) a solid sphere made of tungsten carbide, (c) an aluminum sphere and (d) a solid lucite sphere (bottom).

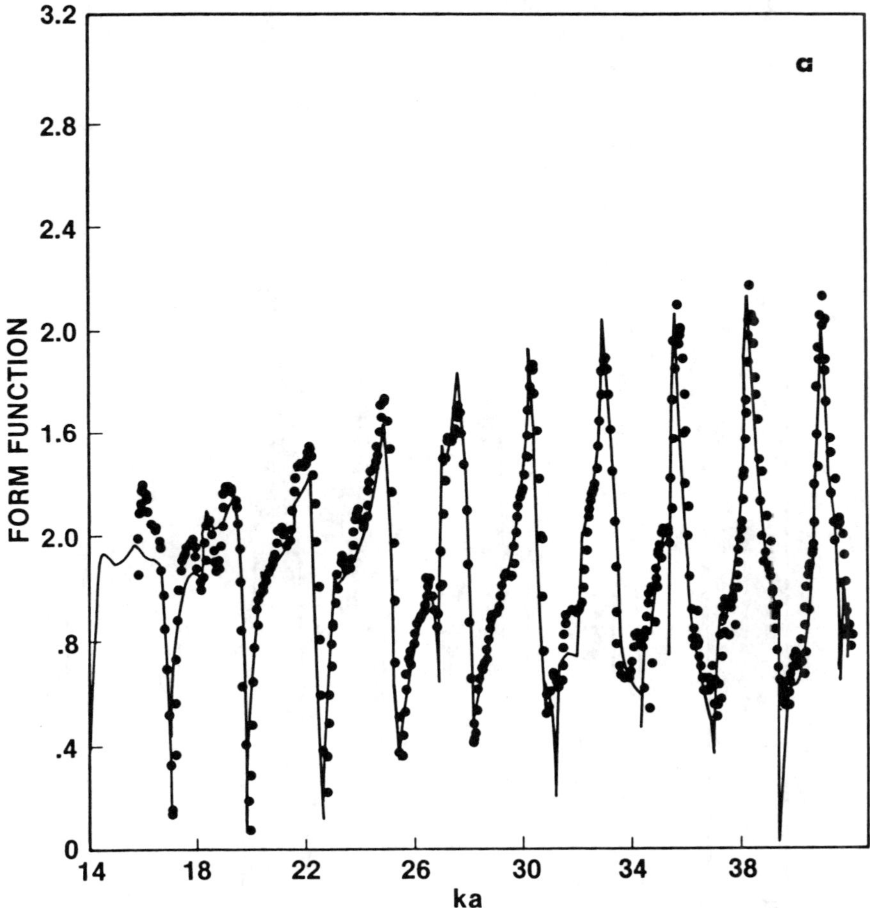


Figure 2(a). Comparison between experimental measurements (points) and normal mode computations (solid curve) of the form function for a tungsten carbide sphere.

the convergence of normal mode series. A number of experiments performed at ARL Texas [12-16], Colorado State University [17] and NRL [18,19] provided impetus for the extension of analytic computations into the high ka region. The rapid development of computer technology, however, made the use of the creeping wave formulation unnecessary from the standpoint of scattering calculations. Practical computations from normal mode series formulations have been made for ka values of over 1000 for spheres and infinite cylinders [20].

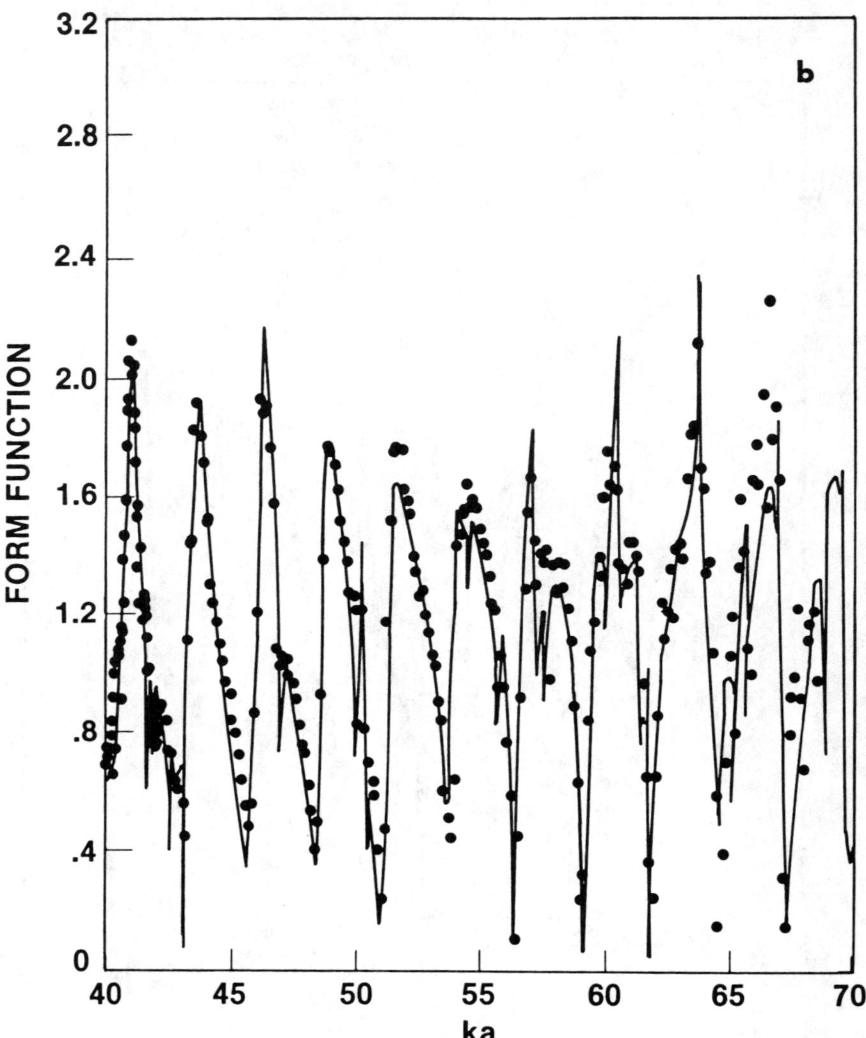


Figure 2(b).

While creeping wave theory became computationally unnecessary, it precipitated the experimental indentification of creeping or circumferential echoes together with the mechanisms responsible for their existence. The time separation of these circumferential echoes can be a valuable classification tool. At frequencies below the excitation of the first elastic resonance, Franz waves or purely geometrically diffracted waves are generated on smooth bodies and travel at approximately the speed of sound in the surrounding medium. Separation between the Franz echo and specular reflection is related to the target dimensions; however, the Franz wave is highly attenuated and of very limited classification value in underwater measurements. Elastic circumferential waves are related to both target dimensions and material constants. These can be separated from one another and from specular reflection at a few limited aspects by interrogating the target with a short incident pulse.

Circumferential waves are especially useful when the target is a thin, hollow, air-filled shell. Elastic circumferential waves isolated in the reflections from a hollow, cylindrical, aluminum shell at beam aspect are shown in Figure 3. The three pictures that comprise Figure 3a are schlieren visualizations of the incident sound pulse, the pulse interacting with the near side of the cylindrical shell and finally the sound field that contains elastic circumferential waves. The schlieren technique is described in several publications [19-21] and offers the distinct advantage of displaying a complete, 360 degree scattering pattern. The hydrophone measurements in Figure 3b show the detection of a series of equally spaced echoes in the back-scattered direction. Backscattered circumferential echoes are displayed in digital form in Figure 3c. The numbered echoes result from the first through seventh transits of the circumferential wave around the cylinder. Because these waves circumnavigate the cylinder, their separation in time yields a measure of the size of the cylinder.

Classification of targets by real time form function correlation is limited, for the most part, to known targets, i.e., to the presence or absence of one particular type of body of known size, shape and material. Further investigation of the form function delineated the relationship between the free body resonances of a body and its form function. The significance of the free body resonances of a submerged target has been quantified through a series of experimental and analytic investigations. Neubauer and Vogt [22] recognized that the deep nulls in the form function for solid elastic spheres were associated with free body resonances. They demonstrated that the ka position at which these nulls occurred could be measured with great accuracy, and that this knowledge could be used to determine the elastic constants of the target material. Measurements of sound speeds to within 1% accuracy have been obtained from scattering measurements on solid spheres and cylinders submerged in a large acoustic pool. Analytic predictions indicate that the measurement technique could be refined to give results an order of magnitude more precise.

a

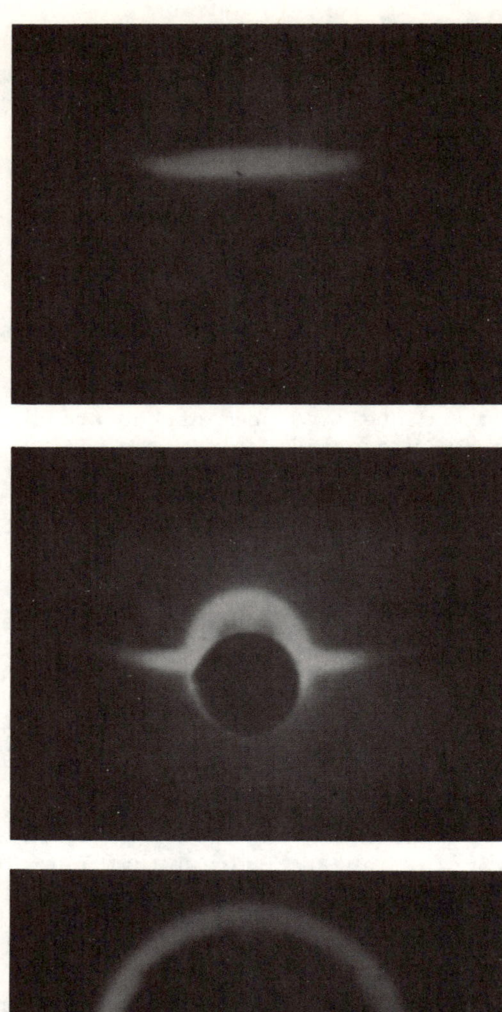

Figure 3(a). Schlieren visualization of the sound scattered by a hollow, cylindrical, aluminum shell showing the incident sound pulse (top), the pulse interacting with the near side of the shell (center) and the sound field including circumferential wave radiation (bottom).

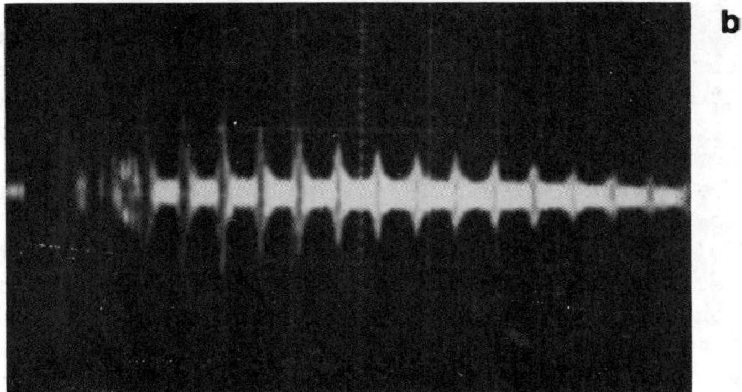


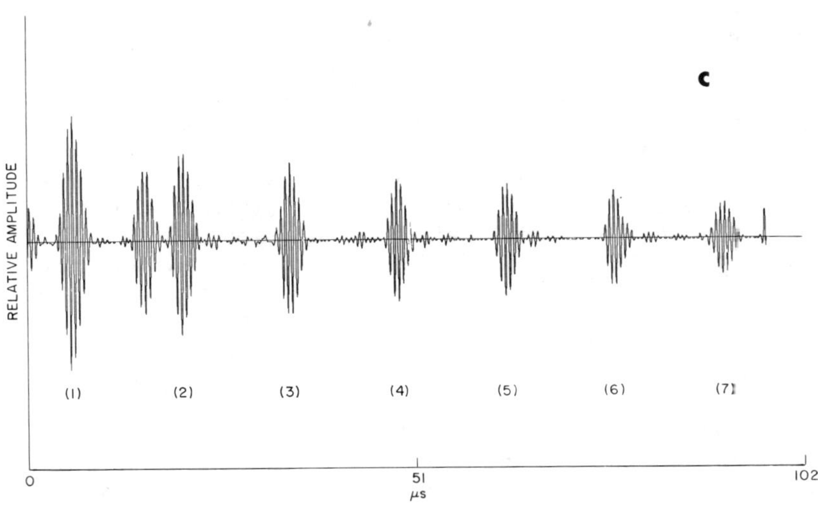


Figure 3(b,c). An oscilloscope trace of the sound scattered by a hollow, aluminum, cylindrical shell (b) followed by a digital representation of the same trace (c).

Mathematical formulation of the resonance behavior of underwater targets was given by Flax et al. [23], who demonstrated that the scattered return from a solid metal target in water could be separated into resonance (elastic) and geometric (rigid) terms. This separation is not simply a mathematical construct, but is a physically correct description of the scattering process. Figure 4a is the form function for an infinitely long aluminum cylinder computed at ka intervals of 0.01. The resonance nulls are identified by the normal mode number n and the eigenvalue number l. Resonances having the same eigenvalue number are related, and when $\ell=1$, the resonances are related to the Rayleigh surface wave. Experimental isolation of the circumferential wave related to the Rayleigh surface wave [24] was made possible by using the information in Figure 4a to determine the appropriate frequency at which it could be observed. The first and second transists of the Rayleigh wave are shown with the specular return in Figure 4b.

The ka position of form function nulls for elastic prolate spheroids were also predicted by simple resonance considerations and verified analytically through T-matrix computations [25].

## EVALUATION OF TIME SIGNATURES AND FORM FUNCTIONS

The elastic behavior of a body has been linked to the observation of circumferential waves in the time signature of the scattered return and to the resonance nulls of the body's measured form function in ka or frequency space. Nevertheless, unambiguous indentification of a particular target still depends upon prescience of the form function of that target. This includes detailed information about its size, shape and material. Considerable difficulty is involved in determining analytically the form function of a finite body of arbitrary shape and material. While the T-matrix formulation of the scattering problem [27] has produced numerical results at small values of ka, computations over a broad ka region remain expensive. Practical solutions to the classification problem depend upon some knowledge of the size, shape, or material of the body in question. In many instances, it is possible to make reasonably accurate assumptions about one or more of these factors. Simple judgements about the nature of the target itself can then be drawn through examination of either the time signature or the measured form function.

A cursory examination of the time signature reveals readily apparent differences between rectilinear and curvilinear bodies as well as between hollow and solid targets. Figure 5a is the echo received from a sound pulse normally incident on an aluminum plate one inch in thickness. Only normal incidence will produce a large specular return from a

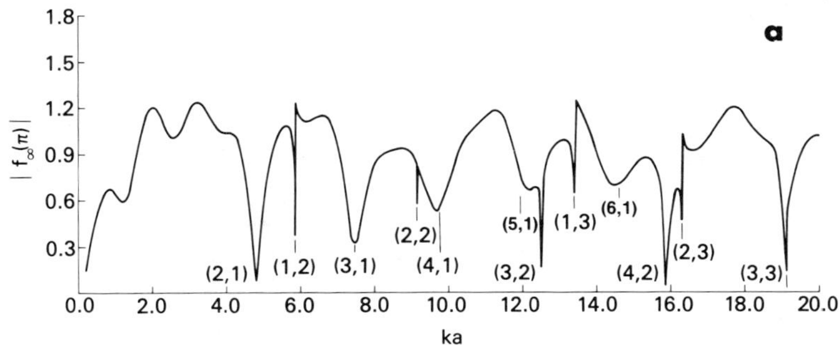


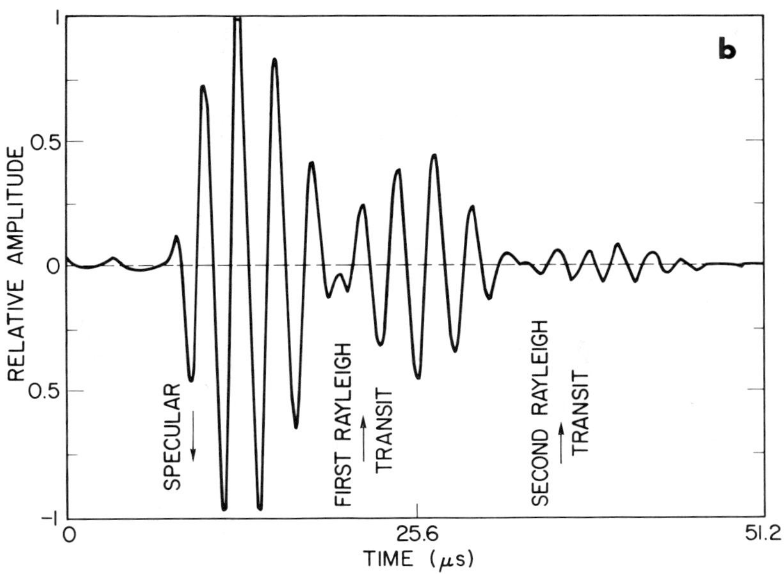


Figure 4(a,b). Form function for an infinitely long aluminum cylinder showing resonance nulls (a) and the time signature with circumferential waves in evidence (b).

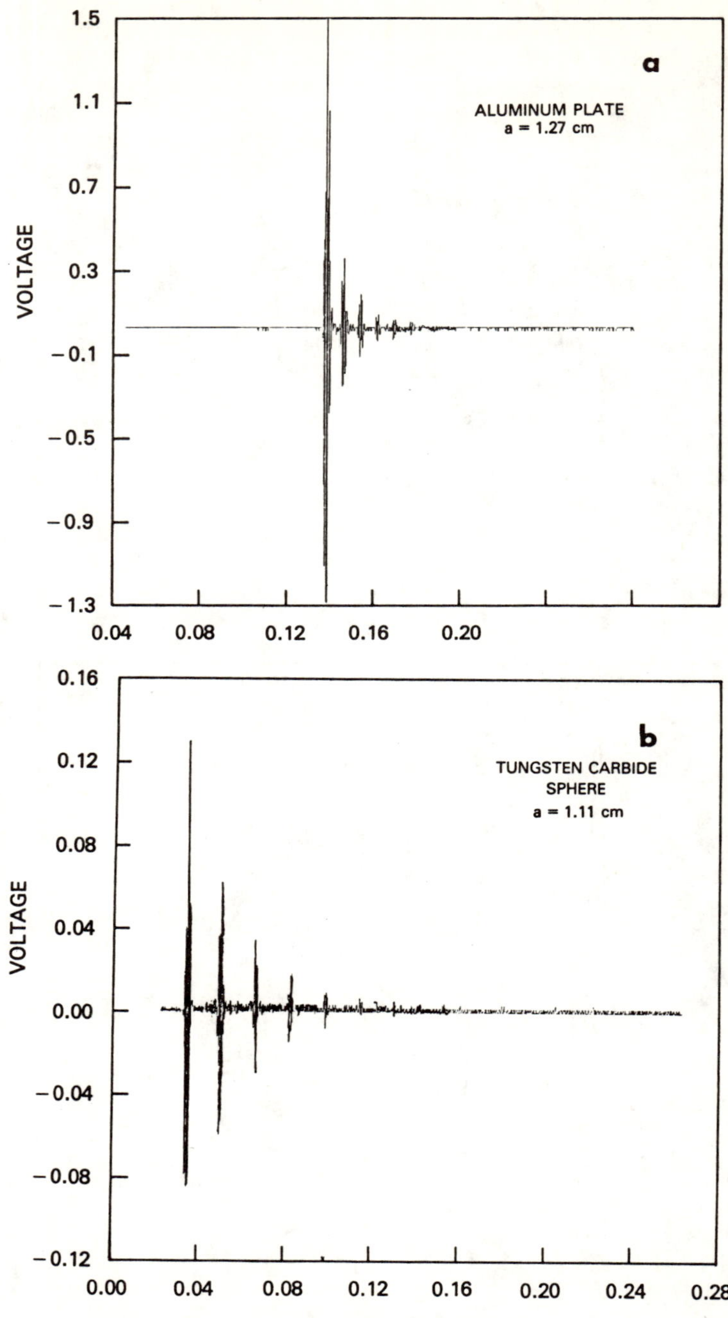


Figure 5(a,b). Backscattered echoes received (a) from a solid aluminum plate roughly milled and (b) from a tungsten carbide sphere 2.22 cm in diameter.

flat surface. The echo from the plate consists of the backscattered return from the front side of the plate, followed by a series of echoes produced by sound penetrating the solid, backscattering off the rear side and reentering the water. This procedure is repeated since there is always some sound reflecting off the solid-liquid interface and some sound penetrating into the medium beyond the interface.

One curvilinear body produces a time signature that, at first glance, might be confused with the plate return. Figure 5b shows the echo received from a solid, tungsten carbide sphere with a 1.11 cm radius. It, too, consists of a specular return followed by a series of equally spaced echoes. The most significant difference between the two figures lies in the size of the returns following the specular. The second echo in the plate signature is down by a factor of 5 in voltage; whereas, the second echo from the sphere is about half the magnitude of the specular return. Only a single path is available for producing the returns in Figure 5a--normal incidence. While a curvilinear body may produce circumferential waves for just one angle of incidence, there are many places on the surface of the body where the sound is incident at precisely that angle. On the sphere, for example, the locus of points at which sound is incident at any single angle is described by a circle. Sound coupling to the surface at all of these points will initiate the same path, thus allowing for a reinforcement or focusing of the sound. The presence of specular return followed by sizeable circumferential echoes is a clear indicator of the presence of a curvilinear body.

The time signature of curvilinear bodies is characterized, especially in the higher ka regions, by the presence of circumferential waves. Neubauer et al. [19] have shown that not all circumferential waves result from sound waves circumnavigating a smooth curved surface. The incident sound may also penetrate the body, be internally reflected along definable paths, undergo mode conversion from shear to longitudinal or the reverse, and emerge as part of an identifiable, smooth wavefront--apparently circumferential. In many cases, the circumferential waves observed in the time signature of a smooth, curved body are dominated by returns that have penetrated the body. For this reason, a direct determination of the radius of a curved body from the separation of the circumferential returns is not possible; however, an analysis of the probable internal paths, similar to that performed by Neubauer [19] allows the extraction of size and shape information whenever the material is known.

Solid targets can be distinguished from shells of similar shape on the basis of their time signatures. The targets selected for this comparison were finite, hemispherically capped cylinders, one solid and one a hollow, air-filled shell. The solid aluminum cylinder is shown in Figure 6a. Its overall length is 1.52 cm, including the hemisphere with a 0.63 cm radius. The dimension chosen for purpose of comparison was half the overall length or 1.26 cm which, for this body, is the length of its cylindrical portion. The cylindrical portion of the brass shell, pictured in Figure 6b, is comparatively shorter--only 20% longer than the radius of the hemisphere; however, the overall length is similar. In the case of the shell, the characteristic dimension, half the overall length, is 1.22 cm. The shell is relatively thin with a ratio of inner to outer radius of 0.9336.

The cylinders were insonified with short pulses centered at 1MHz. The source/target geometry for the measurements is illustrated in Figure 7. With the characteristic dimension, a, chosen as specified, the resultant ka range extended from 30 to 70. The backscattered waveforms in this ka region are markedly different. There are some traces of circumferential waves in the return from the solid cylinder, Figure 8a, but they cannot compare in size or duration with the echo from the brass shell shown in Figure 8b.

The dominant paths for the sound backscattered by the solid cylinder are mode conversion paths inside the material rather than Rayleigh waves circumnavigating the body. The angles at which these paths can be initiated are limited and can be defined for a specific material. The strongly reinforced circumferential waves in the shell's echo result from the generation of Lamb modes in the shell. In this ka region, the first anti-symmetric Lamb mode, or flexural wave, accounts for the large return. The onset of the flexural wave is not only a function of the elastic properties of the material, but is strongly dependent upon the thickness of the shell itself.

Thus for a curvilinear body, there are recognizable features in the time signature that are related to both the material and shape of the body. The echo from the sphere is easily distinguished from that of the hemispherically capped cylinders, even though the sound incident on the cylinder or the shell encounters a spherically shaped body at the chosen aspect. Once the material is known or presumed, the angles at which sound can penetrate a solid body and the shear speed in the material can be used to determine the pathlength through the body from the time separation in the circumferential returns. The separation of circumferential returns can also be used to indicate the possible extent of a hollow body.

The most significant limitation to this type of analysis is that the insonification frequency must be sufficiently high to allow the separation of circumferential returns. The

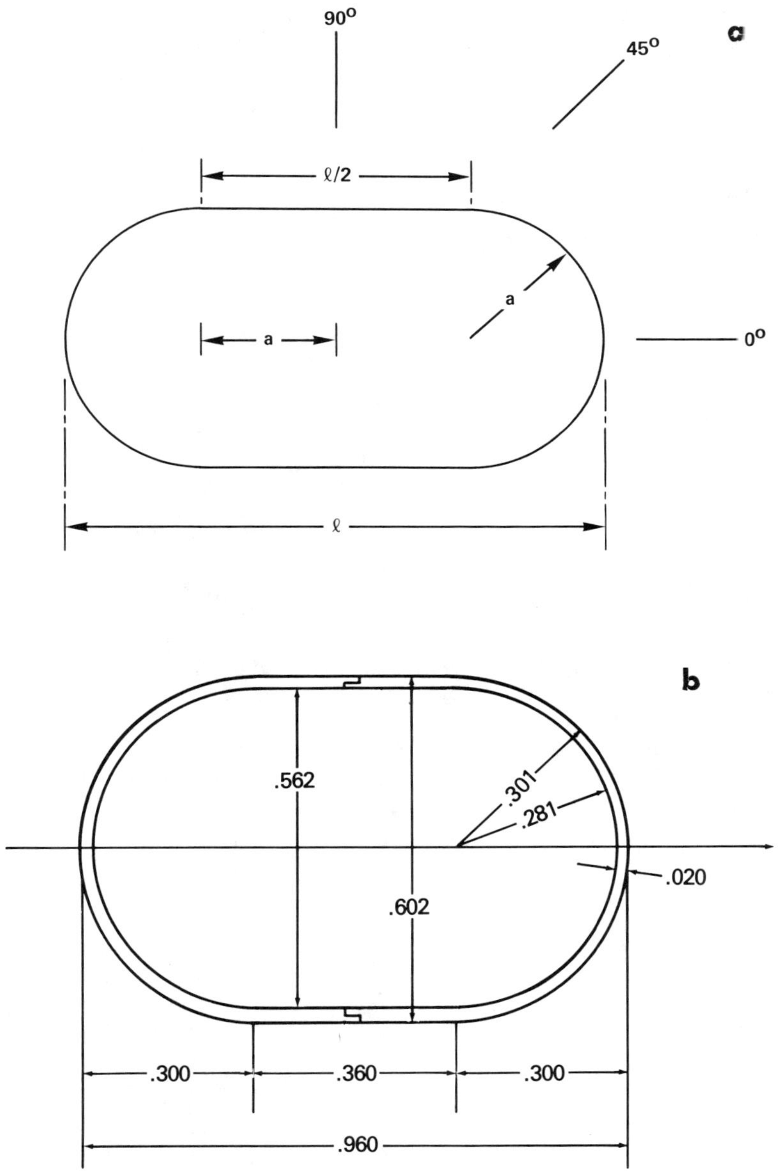


Figure 6(a,b). Drawing of a hemispherically capped cylinder (a) and of a cylindrical shell (b).

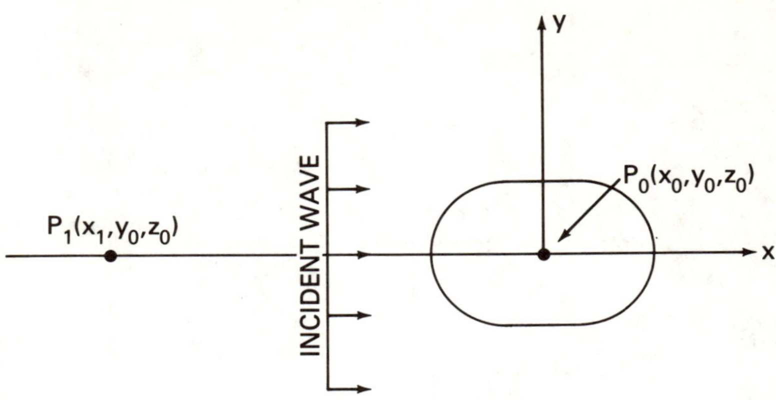


Figure 7. Source/target geometry for acoustic backscattering experiments.

same type of information can be drawn from the form function without this frequency restriction.

The form function is derived experimentally by dividing the Fourier transform of the echo by the Fourier transform on the incident sound pulse. The form functions displayed in Figures 9 and 10 correspond to the time signatures examined in Figures 5 and 8, respectively. The separation of the nulls in the form functions yield a measure of the size of the body, as does the separation of circumferential returns in the time signature. True to its name, the form function bears the imprint of the target's shape.

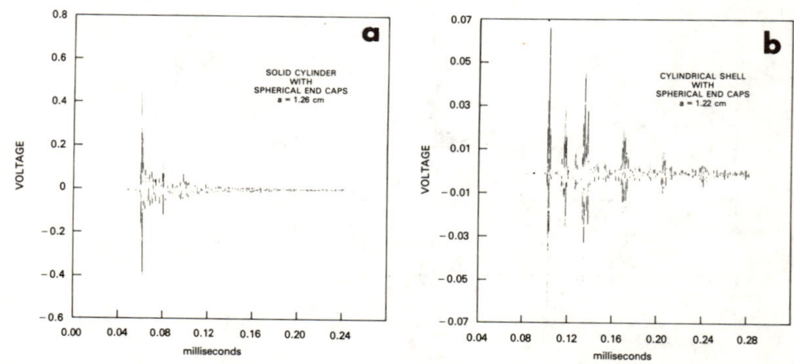


Figure 8(a,b). Backscattered echoes from the solid cylinder (a) and the cylindrical shell (b).

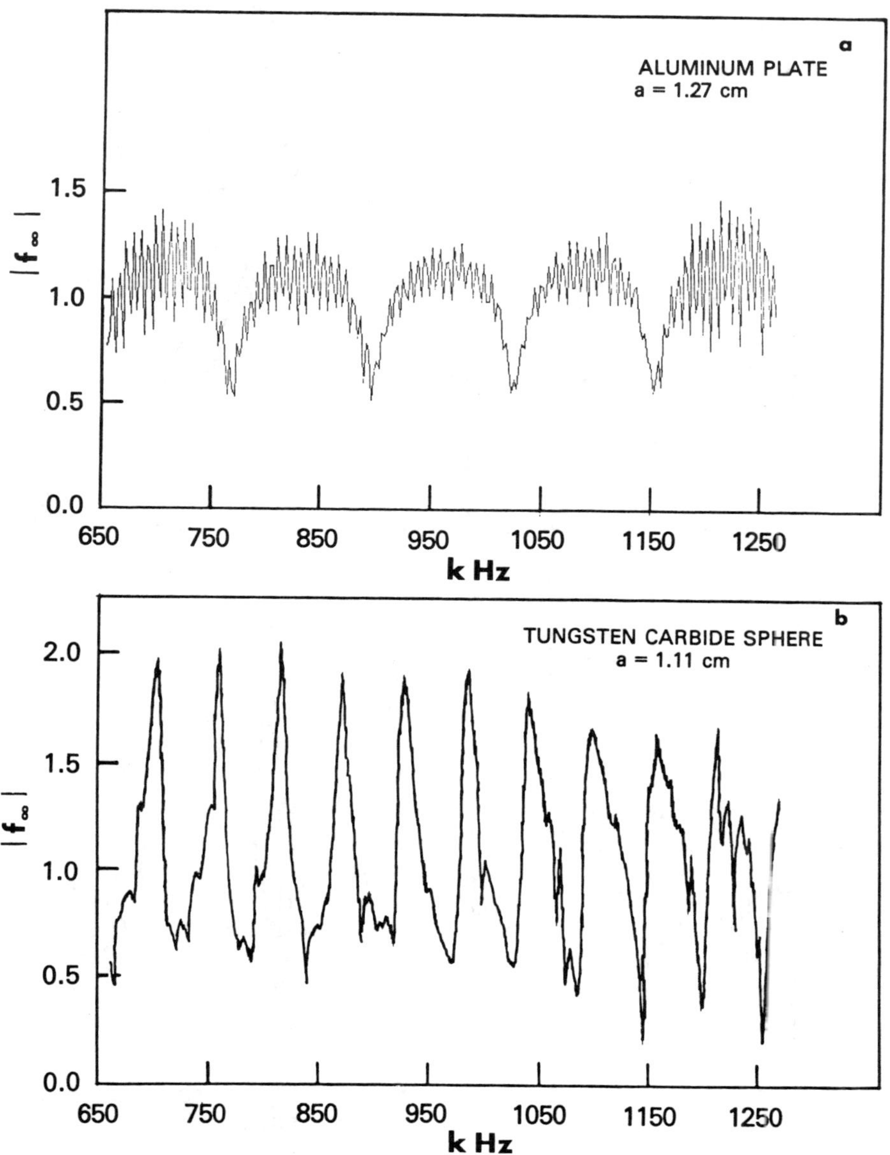


Figure 9(a,b). Form function of a roughly milled aluminum plate 2.54 cm thick (a) and of a tungsten carbide sphere (b).

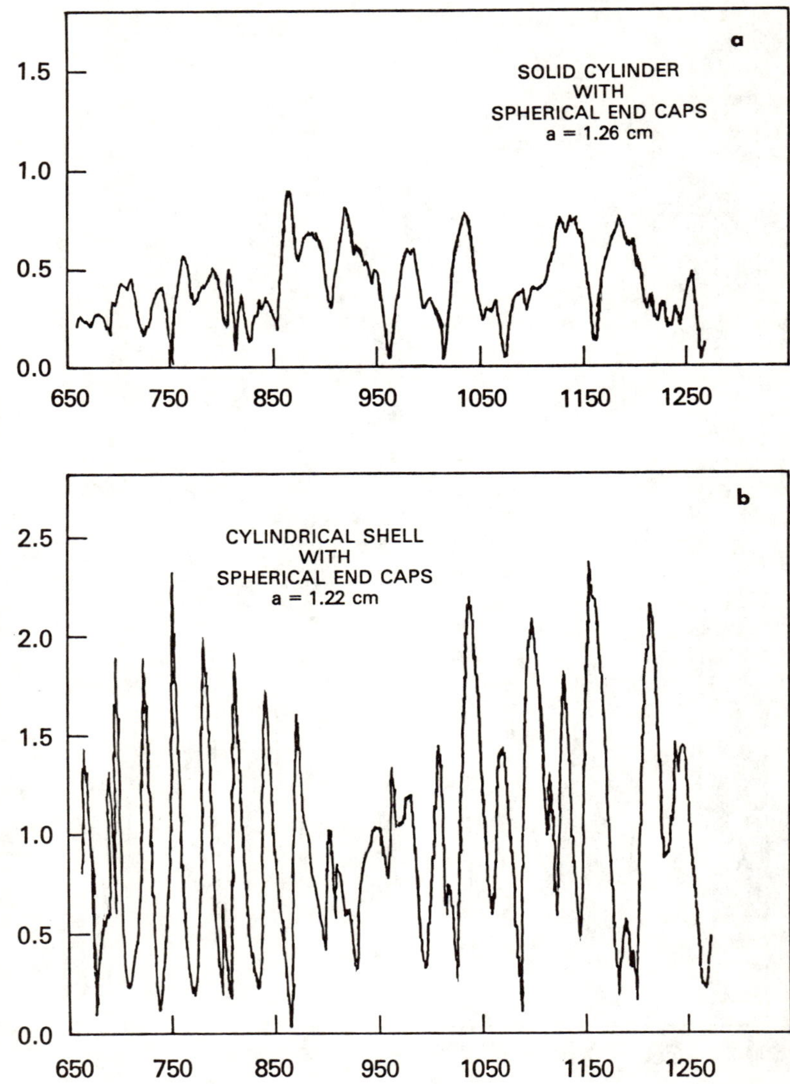


Figure 10(a,b). Form functions at end-on incidence for a hemispherically capped solid cylinder (a) and a cylindrical shell (b).

The differences that could be observed in comparing the time signatures of the plate and sphere are magnified in their form functions. The marks left on the plate by the milling tool produced the rapid oscillations in the form function in Figure 9a. The frequency scale can be converted into the ka scale by multiplying by $2\pi a/c$, where a is the characteristic dimension of the body and c is the speed of sound in water. In determining pathlength, the speed of sound in the material replaces the speed of sound in water. Thus either the characteristic dimension or the speed of sound in the material must be known. The nulls in the form function of the sphere, Figure 9b, correspond to the circumferential waves observed in Figure 5b. Even in lower ka ranges, where the time signature fails to separate the circumferential return, nulls corresponding to the circumferential waves appear in the form function. The onset of the first Rayleigh wave can be observed in the ka region below 10 for most materials and can be used to determine either the material or the size of the sphere.

The form functions for the solid cylinder and the cylindrical shell appear in Figure 10. Again, the cylinder can be distinguished from the shell immediately. While a number of paths contribute to the return from the solid cylinder, careful examination of the nulls in the form function, Figure 10a, reveal a dominant periodicity from which the dominant pathlength can be estimated. The flexural wave in the cylindrical shell causes the dramatic variations in the magnitude of its form function in Figure 10b. This behavior is characteristic of hollow bodies. Since the onset of the flexural wave is strongly dependent on thickness, observation of the form function in the ka region where this behavior begins can yield information about the thickness of the shell.

In all of the comparisons involving either the time signature or the form function, a common problem arises in any attempt to revolve the size and shape of the target--the inability to separate the elastic properties from the size and shape parameters. Pathlength is determined by time separation or null spacing. In both cases the quantity involves the elastic properties under the guise of sound speed as well as the size. Before the problem of target indentification can be moved beyond the level of a complex, albeit reasonable and accurate, guessing game, a method must be found to separate the elasticity from the size and shape parameters.

The resonance theory indicates that the separation between geometric and elastic scattering is possible for both solid and hollow metal targets in water. In some cases this separation can be achieved in the echo itself. The reflection of a high frequency pulse from a tungsten carbide sphere shown in Figure 11 is an example of this separation. The initial backscattered echo is the rigid body return which is

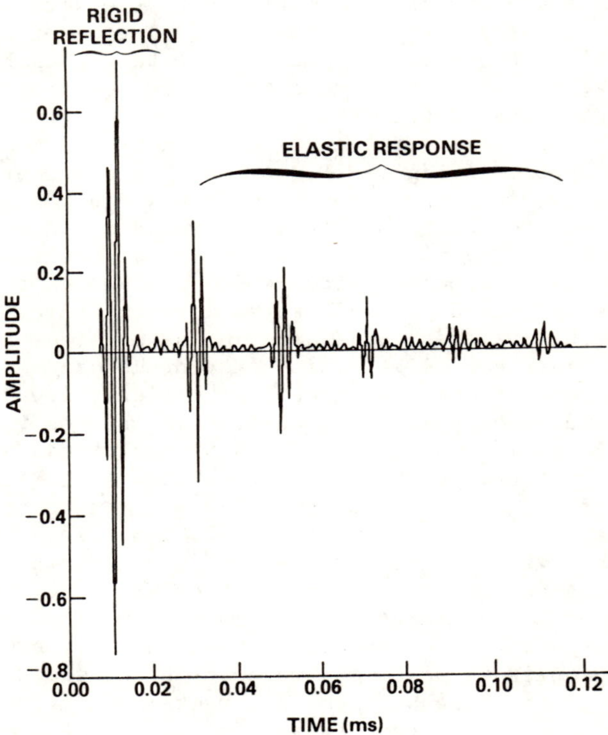


Figure 11. Backscattered echo from a tungsten carbide sphere showing the separation of the rigid and elastic responses.

followed by a series of periodically spaced elastic echoes. The elastic response marked in Figure 11 results from sound penetrating the sphere, undergoing mode conversion and emerging as a circumferential wave. The geometric or rigid reflection is a replica in frequency content and amplitude of a truly rigid sphere [26].

Most practical combinations of target and interrogating pulse do not give a situation as simple as that in Figure 11. For classification of unknown targets, however, it is essential that the geometric and elastic response be separated, even if the separation cannot be performed in the time signature. A major effort is being carried on at NRL to effect this separation and to provide comparisons that will yield size and shape information for unknown targets.

Before any comparisons can be made, an analytic model capable of predicting the return from finite bodies of arbitrary shape must be found. The T-matrix formulation of the scattering problem [27] has resulted in accurate computations of the sound scattered by selected finite, elastic bodies [28]. The finite, hemispherically capped cylinder described in Figure 6a was used for comparison with the T-matrix computations. A single frequency, translated into ka space as k(1/2)= 6, was used in computing the elastic form function. The measurements were taken at 1 degree intervals, rotating the target about the y-axis shown in Figure 7. The results of the comparison are shown at the top of Figure 12. The illustration on the left with the cylinder pictured in the center is the experimentally determined beam pattern. The computed beam pattern is virtually indentical.

In an analytical model, the imposition of rigid boundary conditions simplifies the computational problem. The T-matrix method, which produced such excellent agreement with experimental results for a finite elastic body, should provide accurate beam patterns for rigid bodies of arbitrary shape. In fact, the simplifications introduced by rigid boundary conditions make evaluations at higher values of ka realizable.

The experimentalist cannot turn an elastic body into a rigid body as readily. Resonance theory asserts that the complex return from an elastic body is the superposition of the elastic response on a rigid body return [23]. If this is an accurate description of the physical process, then there must be a means of extracting the rigid body response from the scattered return. At high values of ka where the specular return is separated from the elastic returns, as is the case in Figure 11, the separation can be effected by experimentally gating out the specular return or by deleting it from the digital record. When the insonification frequency is sufficiently low that the rigid return is buried in the elastic response, integration can be used in the time domain or in frequency space to provide an approximation of the

MEASURED AND COMPUTED BEAM PATTERN FOR SOLID FINITE ALUMINUM CYLINDERS

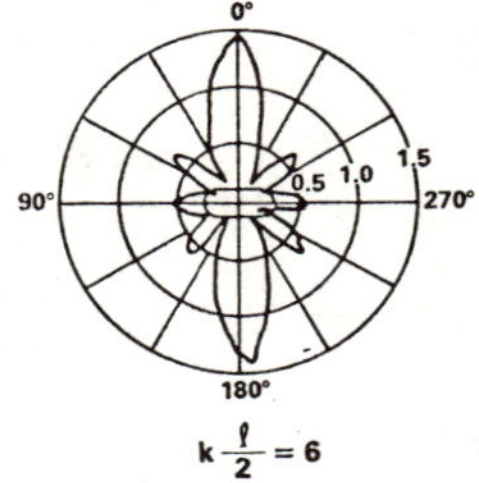


$k\frac{\ell}{2} = 6$

MEASURED INTEGRATED BEAM PATTERN FOR SOLID FINITE CYLINDER

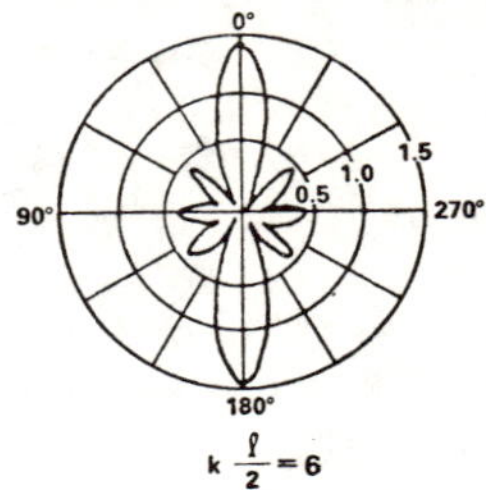


$k\frac{\ell}{2} = 6$

RIGID BODY COMPUTATION

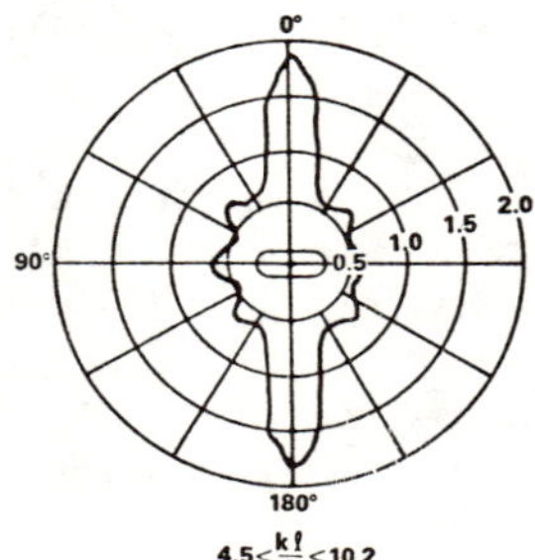


$4.5 < \frac{k\ell}{2} < 10.2$

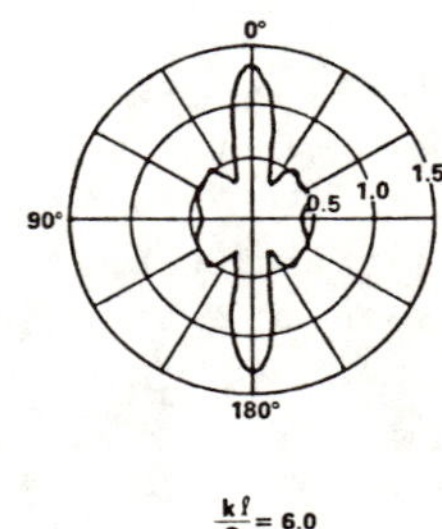


$\frac{k\ell}{2} = 6.0$

Figure 12. Measured and computed beam patterns for a solid, finite aluminum cylinder with complete elastic response (top) and rigid return (bottom).

rigid body response. Integration in the frequency domain involves additional programming, but it produces more accurate results in cases where noise is significant by restricting the computation to only those frequencies present in the incident sound.

The approximate rigid body response for the finite aluminum cylinder shown in Figure 6a was determined using integration in the time domain. The pulse measurements were centered about 0.5MHz, a frequency region in which temporal separation of the specular return was impossible. The ka space spanned extended from 4.5 to 10.2 and included the individual elastic return displayed at the top left of

Figure 12. The analytical rigid body computation was evaluated at 6.0 using the T-matrix. The results are compared at the bottom of Figure 12 with the measured response on the left. The beam patterns are very much the same, but there are differences in magnitude. Time domain investigation was chosen for the comparison because it is the simpler and less accurate method and would provide a "worst case" example. Contributions to the magnitude of the experimental beam pattern are made by noise that lies outside the bandwith of the interrogating sound. Integration in the frequency domain would, therefore, reduce the level of the beam pattern by eliminating this noise.

Elastic effects in hollow bodies are far more apparent in both the time signature, Figure 8, and in the form function, Figure 10. The beam patterns in Figure 13 were obtained by selecting four individual frequencies from a series of broadband pulse measurements taken at 1 degree intervals as the brass shell was rotated about the y-axis shown in Figure 7. There are dramatic variations between any two of these patterns. Integration in the time domain was used to approximate the rigid body response in the ka region that encompassed all of these beam patterns. The result is shown in Figure 14 to the right of the rigid beam pattern calculated with the T-matrix formulation. Even for this case, in which the elastic effects are quite large, the agreement is good.

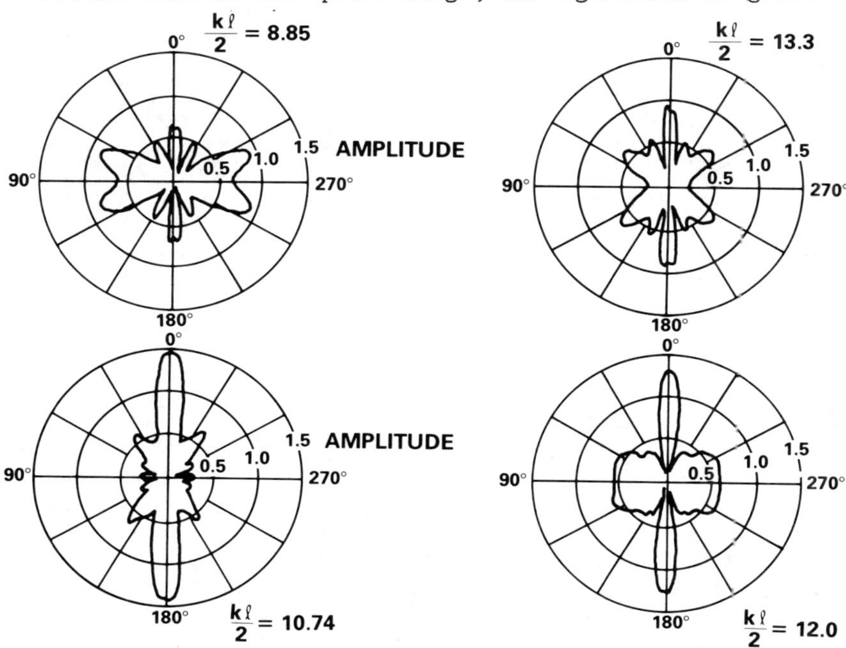


Figure 13. Measured beam patterns for an air-filled, finite cylindrical shell.

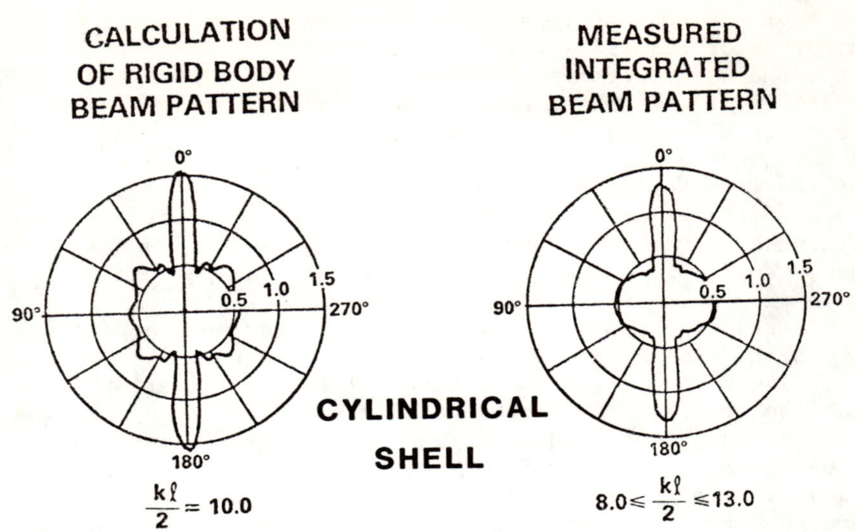


Figure 14. Comparison of the theoretical (left) and experimentally derived rigid response for a cylindrical shell.

Construction of a beam pattern is not always possible; however, the approximate rigid body response from a single echo can add significant information that is already available. The time signature or the form function should determine whether the object is curvilinear or flat. In addition, the form function can describe the body as fundamentally spherical or cylindrical, solid or hollow. The rigid body response derived from the time integrated echo can be compared with well known target strength formulas for rigid cylinders and spheres to determine the possible size of the target [29].

## CONCLUDING REMARKS

Considerable analytical and experimental effort has been devoted to delineating and understanding the mechanisms involved in elastic scattering. Areas of concentration have moved from scattering by simple rigid bodies to the study of elastic effects in simple shapes. Current analytical models show promise of providing details of the scattered response of finite bodies of arbitrary shape and size. Digital processing techniques have kept pace with theoretical developments, permitted real time processing of broadband signals and correlation with stored form functions. By using transient pulse analysis methods, the experimentalist can derive an approximate rigid body response from a single echo produced by an elastic body. With these advances the groundwork has been laid for developing an accurate method that can be used to classify submerged bodies by acoustic means.

REFERENCES


1. Faran, J. J., Jr. "Sound Scattering by Solid Cylinders and Spheres," *J. Acoust. Soc. Am.* 23(4):405-418 (1951).

2. Faran, J. J., Jr. "Sound Scattering by Solid Cylinders and Spheres," ONR Technical Memorandum No. 22 (Cambridge: Harvard University, 1951).

3. Hickling, R. "Analysis of Echoes from a Solid Elastic Sphere in Water," *J. Acoust. Soc. Am.* 34(10):1582-1592 (1962).

4. Hickling, R. "Analysis of Echoes from a Hollow Metallic Sphere in Water," *J. Acoust. Soc. Am.* 36(6):1124-1137 (1964).

5. Neubauer, W. G., Vogt, R. H. and Dragonette, L. R. "Acoustic Reflection from Elastic Spheres. I. Steady-State Signals," *J. Acoust. Soc. Am.* 55(6):1123-1129 (1974).

6. Dragonette, L. R., Vogt, R. H. and Neubauer, W. G. "Acoustic Reflection from Elastic Spheres and Rigid Spheres and Spheroids. II. Transient Analysis," *J. Acoust. Soc. Am.* 55(6):1130-1137 (1974).

7. Dardy, H. D., Bucaro, J. A., Schuetz, L. S. and Dragonette, L. R. "Dynamic Wide-Bandwidth Acoustic Form-Function Determination," *J. Acoust. Soc. Am.* 62(6):1373-1376 (1977).

8. Doolittle, R. D. and Überall, H. "Sound Scattering by Elastic Cylinders," *J. Acoust. Soc. Am.* 43(1):1-14 (1968).

9. Ugincius, P. and Überall, H. "Creeping-Wave Analysis of Acoustic Scattering by Elastic Cylindrical Shells," *J. Acoust. Soc. Am.* 43(5):1025-1035 (1968).

10. Frisk, G. V. and Überall, H. "Creeping Waves and Lateral Waves in Acoustic Scattering by Large Elastic Cylinders," *J. Acoust. Soc. Am.* 59(1):46-54 (1976).

11. Watson, G. N. "The Diffraction of Electric Waves by the Earth," *Proc. Roy. Soc. (London)* A95:83-99 (1919).

12. Barnard, G. R. and McKinney, C. M. "Scattering of Acoustic Energy by Solid and Air-Filled Cylinders in Water," *J. Acoust. Soc. Am.* 33(2):226-238 (1961).

13. Diercks, K. J., Goldsberry, T. G. and Horton, C. W. "Circumferential Waves in Thin-Walled Air-Filled Cylinders in Water," *J. Acoust. Soc. Am.* 35(1):59-64 (1963).

14. Horton, C. W., King, W. R. and Diercks, K. J. "Theoretical Analysis of the Scattering of Short Acoustic Pulses by a Thin-Walled Metallic Cylinder in Water," *J. Acoust. Soc. Am.* 34(12):1929-1932 (1962).

15. Goldsberry, T. G. "Reflection of Circumferential Waves from a Slit in a Thin-Walled Cylinder," *J. Acoust. Soc. Am.* 42(6):1298-1305 (1967).

16. Horton, C. W. and Mechler, M. V. "Circumferential Waves in a Thin-Walled Air-Filled Cylinder in a Water Medium," *J. Acoust. Soc. Am.* 51(1):295-303 (1972).

17. Bunney, R. E., Goodman, R. R. and Marshall, S. W. "Rayleigh and Lamb Waves on Cylinders," *J. Acoust. Soc. Am.* 46(5):1223-1233 (1969).

18. Neubauer, W. G. "Experimental Measurement of 'Creeping' Waves on Solid Aluminum Cylinders in Water Using Pulses," *J. Acoust. Soc. Am.* 44(1):298-299 (1968).

19. Neubauer, W. G. and Dragonette, L. R. "Observation of Waves Radiated from Circular Cylinders Caused by an Incident Pulse," *J. Acoust. Soc. Am.* 48(5):1135-1149 (1970).

20. Flax, L. "High ka Scattering of Elastic Cylinders and Spheres," *J. Acoust. Soc. Am.* 62(6):1502-1503 (1977).

21. Neubauer, W. G. and Dragonette, L. R. "Diffraction of a Sound Pulse by a Hollow Cylinder in Water," *J. Acoust. Soc. Am.* 48(1):101 (1970).

22. Vogt, R. H. and Neubauer, W. G. "Relationship Between Acoustic Reflection and Vibrational Modes of Elastic Spheres," *J. Acoust. Soc. Am.* 60(1):15-22 (1976).

23. Flax, L., Dragonette, L. R. and Uberall, H. "Theory of Elastic Resonance Excitation by Sound Scattering," *J. Acoust. Soc. Am.* 63(3):723-731 (1978).

24. Dragonette, L. R. "The Influence of the Rayleigh Surface Wave on the Backscattering by Submerged Aluminum Cylinders," *J. Acoust. Soc. Am.* 65(5):1570-1572 (1979).

25. Flax, L., Dragonette, L. R., Varadan, V. K. and Varadan, V. V. "Analysis and Computation of the Acoustic Scattering by an Elastic Prolate Spheroid Obtained From the T-Matrix Formulation," (Submitted for publication).

26. Dragonette, L. R., Numrich, S. K. and Frank, L. J. "Calibration Technique for Acoustic Scattering Measurements," *J. Acoust. Soc. Am.* 69(4):1186-1189 (1981).

27. Varadan, V. K. and Varadan, V. V., Eds. *Acoustic, Electromagnetic and Elastic Wave Scattering - Focus on the T-Matrix Approach* (New York: Pergamon Press, 1980).

28. Numrich, S. K., Varadan, V. V. and Varadan, V. K. "Scattering of Acoustic Waves by a Finite Elastic Cylinder Immersed in Water," (Submitted for publication).

29. Urich, R. J. *Principles of Underwater Sound for Engineers* (New York: McGraw-Hill, Inc., 1967), p. 266.

CHAPTER 10

# AN OVERVIEW OF ULTRASONIC SPECTROSCOPY APPLIED TO NONDESTRUCTIVE MATERIALS EVALUATION

*Dale W. Fitting and Laszlo Adler*
Department of Welding Engineering
The Ohio State University
Columbus, Ohio 43210

Ultrasonic spectroscopy is the study of ultrasonic waves resolved into their Fourier frequency components. Since, in general, all ultrasonic wave interaction with material structure is frequency dependent, ultrasonic spectroscopy provides a promising method for nondestructive materials evaluation. Ultrasonic spectroscopy enhances conventional amplitude A-scan inspection by using broadband pulses and performing frequency analysis to obtain a spectral signature of the material in question.

In an analogy with light, the specimen is insonified with 'white sound.' Defects or material structure alter the 'color' of the scattered field. Ultrasonic spectroscopy has recently been applied to a variety of N.D.E. problems: with most of the attention directed toward flaw characterization, that is, determining geometrical and surface characteristics of flaws. Additionally, the technique has been used to replace mechanical profilometer measurement of surface texture (periodic and randomly rough). A good deal of attenuation has been given to the assessment of the adhesive and cohesive strength of adhesively bonded joints. Ultrasonic spectroscopy is also being utilized to determine strength-related properties such as interlaminar shear strength of composite materials. Empirical correlations have evolved between the frequency dependence of ultrasonic attenuation and fracture toughness for several types of steels. Because the interaction of elastic waves and a polycrystalline material depends on the ratio of wavelength to grain dimensions, microstructure (grain size distribution) is amenable to study by ultrasonic spectroscopic techniques. Additionally, long wavelength scattering information may be used to estimate fracture mechanics parameters for a specimen.

Each of these uses of ultrasonic spectroscopy has arisen because a definite relation (theoretical or experimental) has been found between the frequency response and an important material characteristic.

## INTRODUCTION

Ultrasonic Spectroscopy is the frequency domain study of ultrasonic wave interactions with a material. Ultrasonic spectroscopy proves valuable because many material properties are manifested as amplitude or phase changes in the ultrasonic waves interrogating the specimen.

Initial material characterization studies were carried out utilizing narrowband ultrasonic pulses. Gericke [1] pioneered the use of broadband pulses for material property investigation. Spectrum analysis of the pulse yielded information over a range of frequencies. He also appears to have been the first to use the term "ultrasonic spectroscopy". In an analogy with light, the specimen was insonified with 'white sound'. Defects or material structure would alter the 'color' of the scattered waves.

Other investigators refined ultrasonic spectroscopic instrumentation and broadened the list of applications [2]. In addition to the use of spectroscopy for defect characterization and assessment of adhesive bonds, it has proven useful for: determining surface properties and subsurface gradient, monitoring corrosion, deducing strength-related material properties, investigating the microstructure of a specimen as well as the measurement of frequency-dependent attenuation and velocity.

## ULTRASONIC SPECTROSCOPIC SYSTEMS

All ultrasonic spectrum analysis systems contain provisions for (1) generating ultrasound, (2) receiving a portion of the ultrasound which has interacted with the material under study, and (3) analyzing the received wave to determine the magnitude and phase of the signal over a range of frequencies. In a common system configuration (Figure 1), an

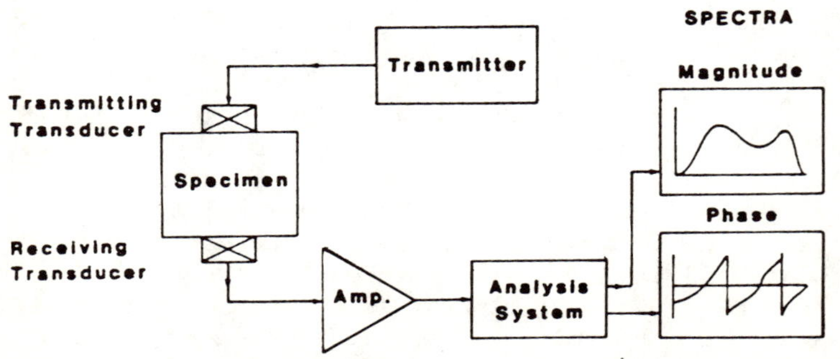


Figure 1. Generalized ultrasonic spectroscopy system.

electrical signal generated by the transmitter is applied to the transmitting transducer. Electrical energy is converted to mechanical energy by the transducer, producing an ultrasonic wave. Interactions of the wave with the material alter its amplitude, phase and direction. A receiving transducer intercepts part of the energy and converts the ultrasonic energy back to an electrical signal. Because the signal is usually small, an amplifier is used to boost its amplitude. An analysis system sorts out the amplitude and timing of the ultrasonic interactions to present magnitude and phase spectra.

Since experimental antifacts must be removed from the displayed spectra, it has been proposed [3] the spectroscopic instrument be modelled as a linear, time-invariant (LTI) system. Components can be analyzed in detail and their interaction to form the instrument may be determined. The behavior of a LTI system is described by its impulse response (in the time domain) or its frequency response (in the frequency domain). The two descriptions of system response are equivalent and are related by the Fourier transform pair.

Importantly, if the input [I(f)] and output [O(f)] from the specimen can be determined, then the ultrasonic transfer function of the material h(t) [or H(f)] may be determined by deconvolution.

Figure 2 illustrates an ultrasonic spectroscope modelled as a LTI system. Notice that for most of the components, the signal of interest is an electrical waveform. However, for an important section of the system the relevant signal is an ultrasonic wave propagating through the coupling media and material under study. To evaluate the material transfer function, the characteristics of each of the system components must be determined, and its effect removed by deconvolution.

Ideally, components should exhibit an amplitude and phase response which is uniform for a wide range of frequencies. Piezoelectric transmitting and receiving transducers and often the mechanical coupling layers have transfer functions which are notoriously modulated. Although the nonuniformities may be compensated for in the deconvolution processing step, signal-to-noise ratio varies across the spectrum.

As conventionally operated, a piezoelectric transducer is a mechanical resonant system. Modification of the boundary conditions (loading) may be utilized to broaden the response (Figure 3). Although the modulated response is undesirable, the high transduction efficiency of the piezoelectric ceramics accounts for their popularity. Alternate means of transduction include electrostatic (capacitive) and electromagnetic devices as well as laser-generated ultrasonic waves.

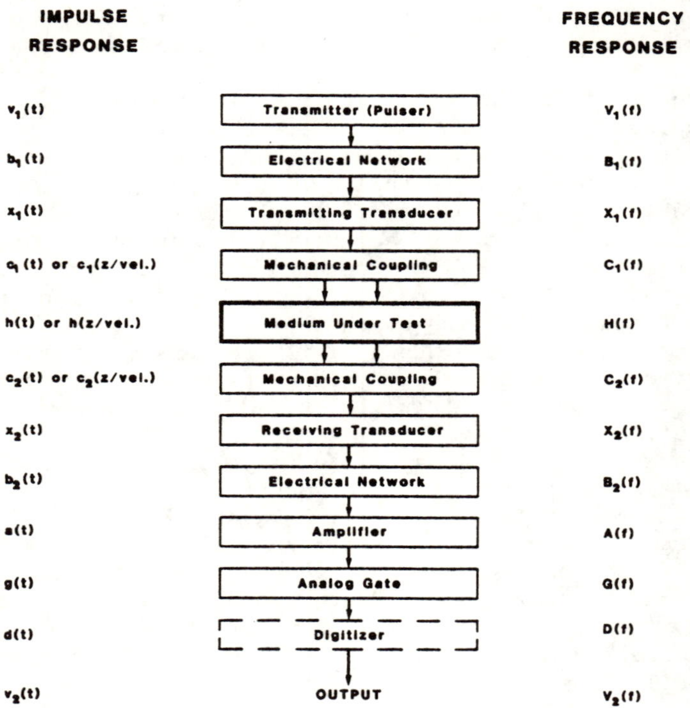


Figure 2. Elements of an Ultrasonic Spectroscopic System Modelled as a Linear, Time-Invariant System.

The radiation field of ultrasonic energy from the transducer into the sample is not ideal. Rather than emitting a cylindrical or rectangular beam, transducers create a field which diverges in lobes and so decreases with distance (Figure 4). These diffraction effects are, of course, related to the ratio of wavelength to transducer dimensions. Thus for a given transducer, divergence changes with frequency (Figure 5). In our LTI system analysis, compensation for the nonideal effects in the radiation field must be performed.

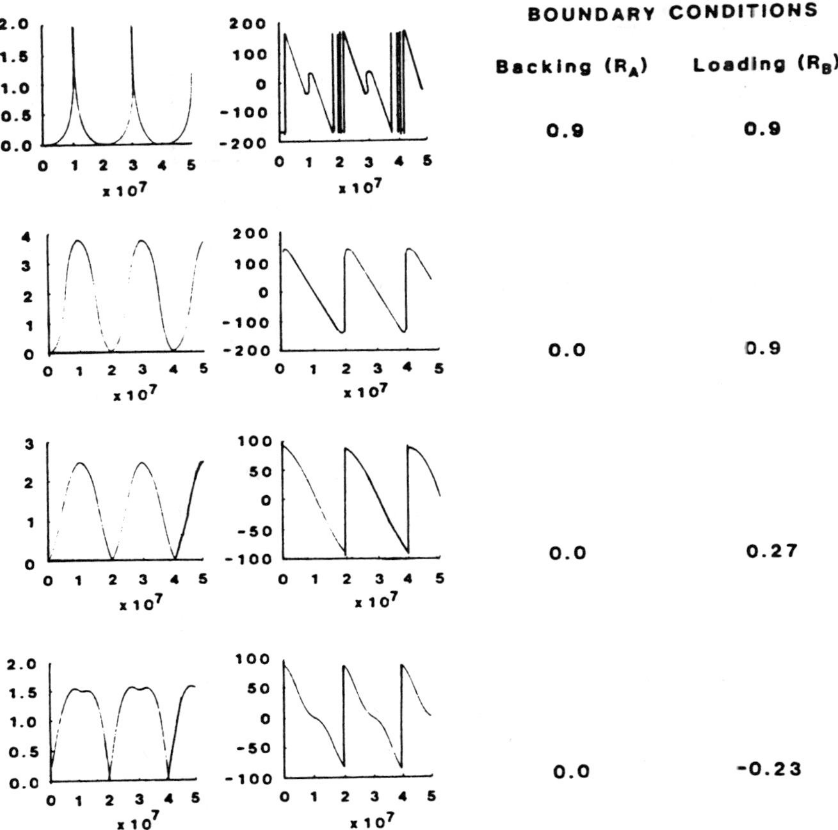


Figure 3. Frequency Response of a Piezoelectric Transducer for a Variety of Loading Conditions.

Assuming we have been able to characterize all components in the spectroscopic system, we move on to the investigation of frequency-dependent ultrasonic wave interactions with a material.

## APPLICATIONS OF ULTRASONIC SPECTROSCOPY

### Defect Characterization

Mathematically, ultrasonic wave scattering from defects is quite complex. Exact theories exist for only a few idealized configurations. A number of approaches to the problem

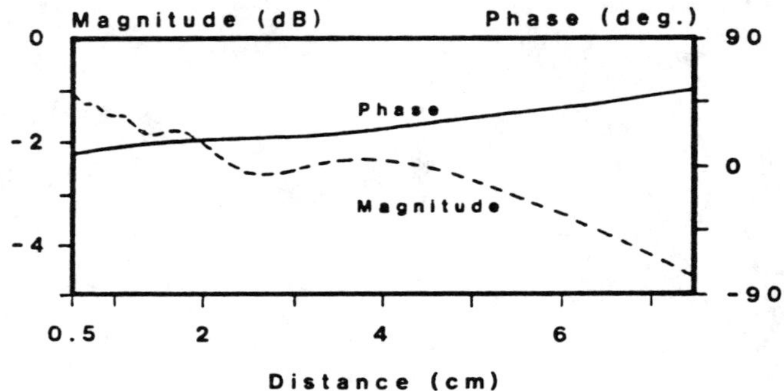


Figure 4. Transfer Function of Radiation Coupling for Fixed Transducer Radius (3.175 mm) and Frequency (2.25 MHz) as a Function of Distance (after Rhyne [4]).

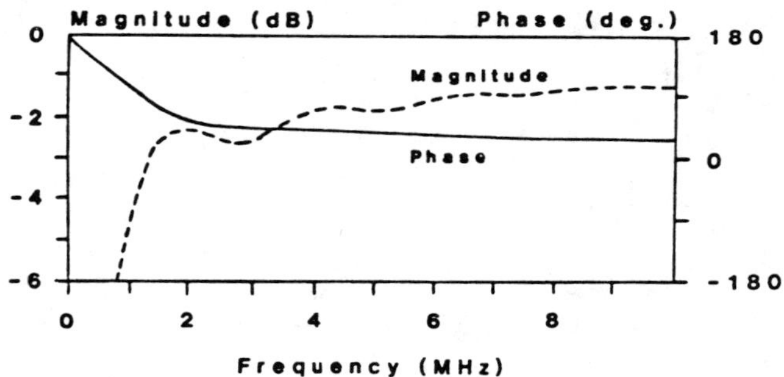


Figure 5. Transfer Function of Radiation Coupling for a Fixed Transducer Radius (3.175 mm) and Distance (30 mm) as a Function of Frequency (after Rhyne [4]).

using a variety of approximate solutions have recently evolved. The scattering problems may be broadly categorized by defect type and ratio of ultrasonic wavelength to defect dimensions. In the former category, scatterers are classified as being bounded by a smooth closed surface (such as spheroids) or by planar surfaces having sharp edges. The second classification is by ka, where k is the wave number and a is the dimension of the discontinuity. In general, a given approximate theory may be valid for a given defect type over only a limited range of ka.

The scattering of ultrasonic waves by a flaw is shown schematically in Figure 6. An incident wave is scattered by the defect and a receiver placed at a point described by polar angle θ and azimuthal angle φ.

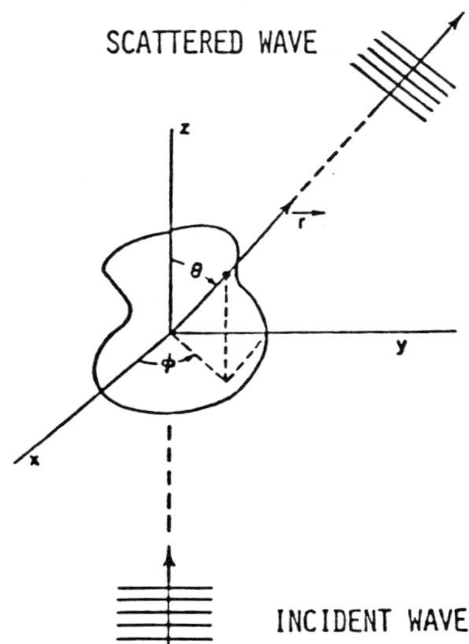


Figure 6. Geometry of Ultrasonic Wave Scattering from a Discontinuity.

In the ARPA/AFML program on Quantitative NDE, diffusion-bonding techniques were utilized to fabricate simulated defects. Sets of titanium samples were prepared having internal defects of well-characterized shapes in a variety of sizes. The artificial flaws are cavities in the shape of spheres, oblate and prolate spheroids, flat-bottomed holes, as well as circular and elliptical cracks.

Tittman and others [5] studied the angular and frequency dependence of scattering from bulk defects by using a contact technique. On Figure 7, representative experimental data are given and theoretical results shown for the frequency dependence of the backscattered power from an oblate spheroid, with 2:1 aspect ratio, at polar angles $10^{o}$ and $60^{o}$.

In Figure 8, experimental and theoretical scattered longitudinal and shear wave amplitude spectra are shown for spherical, oblate and prolate spheroidal cavities. Information concerning size and shape are apparent in both the frequency and angular dependence of the scattered waves.

Additional experiments compared amplitude spectra for waves diffracted from circular cracks to the results of a theoretical analysis using elastodynamic ray theory (Achenbach, et al [7]). For a 2500 micron circular crack, the experimental spectra compared favorably with theory. An

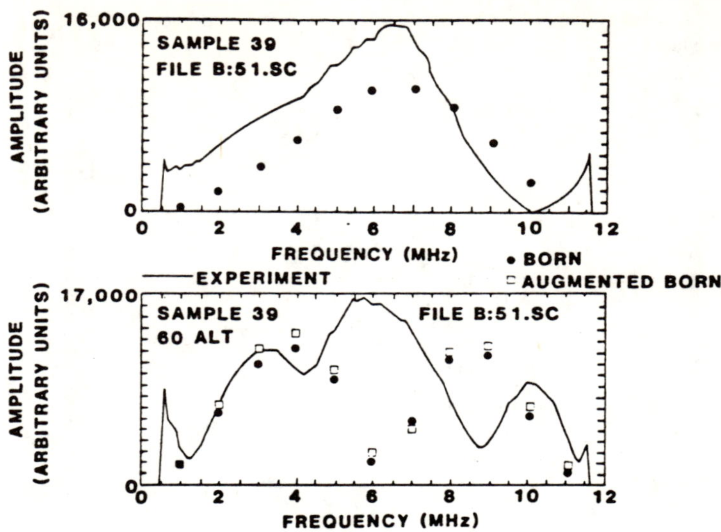


Figure 7. Comparison of Approximate Theories with Data from Backscattering by an Oblate Spheroid for 10° (upper figure) and 60° (lower figure) Off the Axis of Symmetry of the Defect (from Tittmann and Elsley [5]).

extension to the ray theory was used to determine the scattering from a 2500 x 1250 micron elliptical crack. Comparison with experimental spectra is illustrated in Figure 9. Although there are some discrepancies in amplitude, the periodicity in the theoretical and experimental spectra agree favorably.

Sachse and Pao [9,10] studied the frequency dependence of broadband ultrasonic pulses scattered from fluid-filled cavities in a solid. They applied ray acoustics to analyze the various types of waves produced in the cavity. Figure 10 shows the amplitude-time record and power spectra of signals from a water-filled 1/16 inch diameter cylindrical cavity in aluminum.

## Adhesive Bonds

Laminates of metal and polymer (adhesive bonds) have been studied for a number of years using ultrasonic spectroscopic methods. Since the strength of the bond is very important it was hoped that this property could be deduced from nondestructive measurements. A number of models for wave propagation in a multilayer laminate have been developed. Most

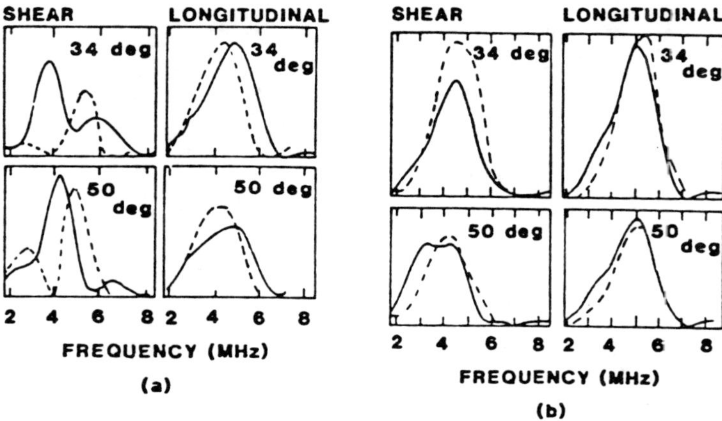


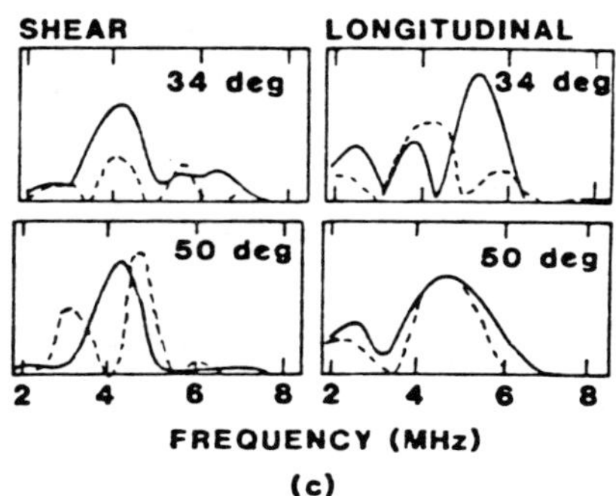


Figure 8. Comparison of Experimental Data to the Born Approximation Theory. Scattered intensity is plotted versus frequency for scattering from (a) 800 micron sphere, (b) 200 x 800 micron oblate spheroid and (c) 1600 x 800 micron oblate spheroid and (c) 1600 x 800 micron prolate spheroid in titanium (from Adler [6]).

experimental investigations demonstrate good agreement with theory. However, these tests were usually performed on highly idealized bonds. More detailed models are required to describe wave propagation in 'real' bonds. Additionally, the strong effects on strength of the lapped edges must be accounted for in any model worthy of serious consideration.

Theoretical investigations of wave propagation in a multilayer laminate by Lloyd [11], Chang, Couchman and Yee [12], Scott and Gordon [13] and Rose and Meyer [14] indicated a resonant behavior of the composite. Figure 11 illustrates

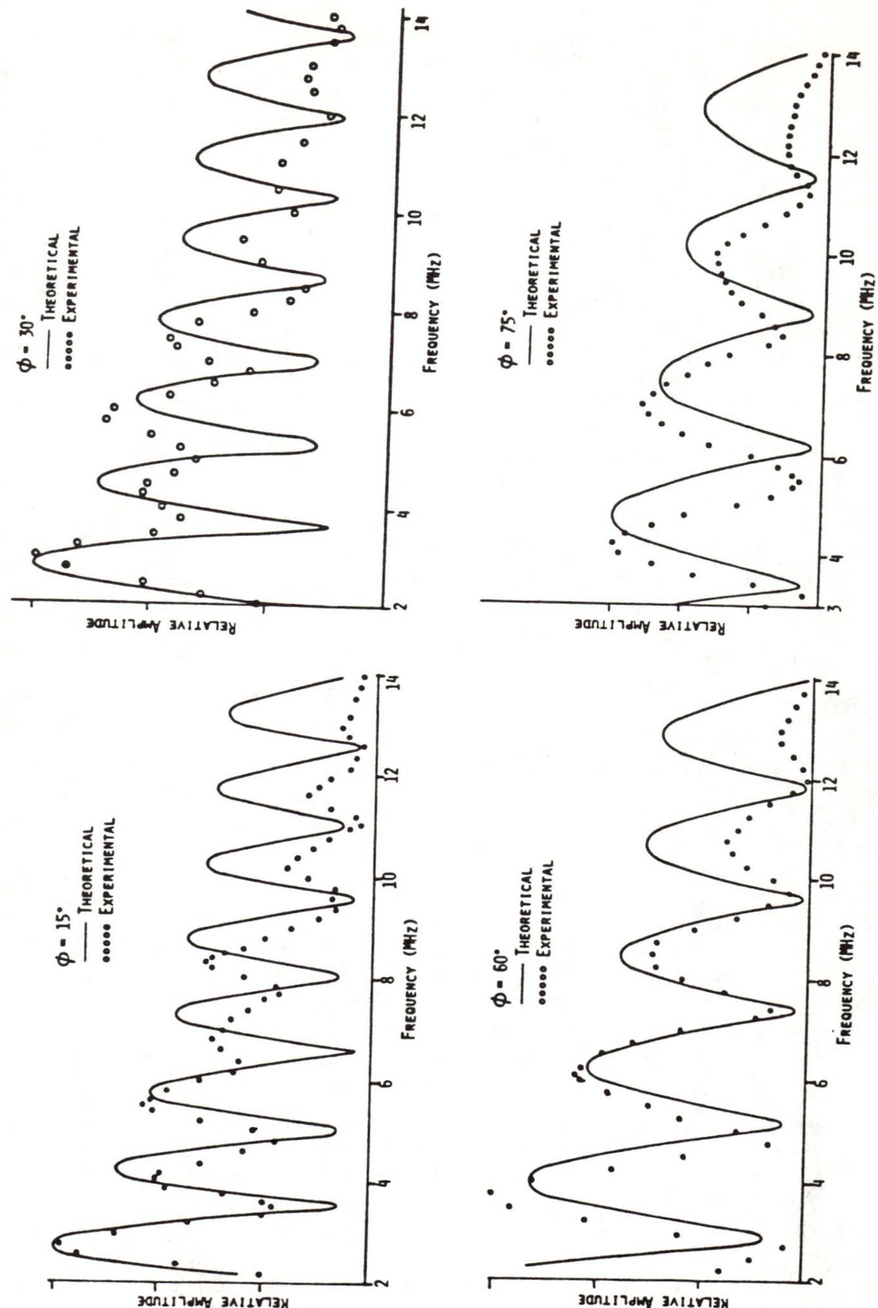


Figure 9. Amplitude Spectra of Scattered L Waves from a 2500 x 1250 micron Elliptical Crack in Titanium Along Different Azimuthal Directions (polar angle = 60°) (from Adler [8]).

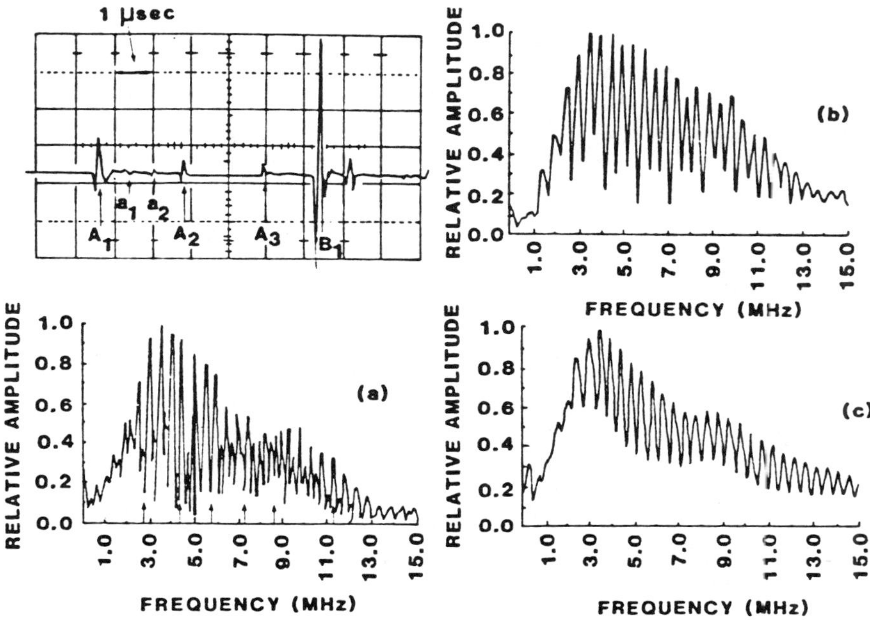


Figure 10. Amplitude-Time Record (a) and Power Spectrum (b) of Waves Scattered from a Water-Filled Cavity. (c) Spectrum of pulses $A_1$, $a_1$, $a_2$ and $A_2$. (d) Spectrum of pulses $A_1$ and $a_2$ (from Pao and Sachse [9]).

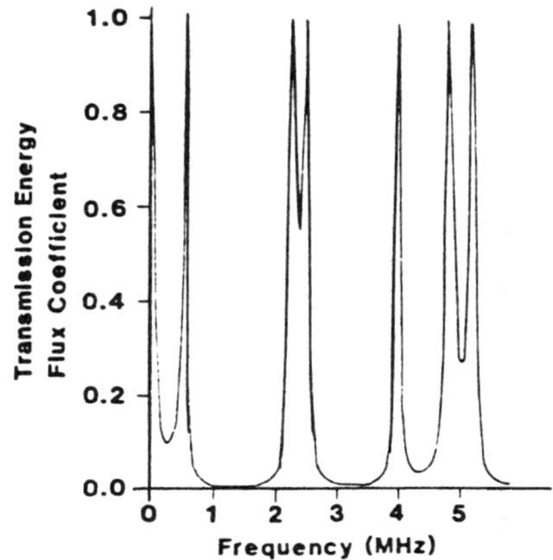


Figure 11. Ultrasonic Transmission Spectrum of a 3-Ply Laminate (Two .05" Al plates bonded by .012" of adhesive) (after Couchman, et al [15]).

the transmission spectrum for a 3-layer laminate. The peaks may be identified with resonance of individual layers, combinations of layers and of the entire structure.

Chang, et al [16] found a correlation between the bandwidth (B) of the antiresonances and the shear strength of step-lap bond specimens. Alers, et al [17] performed a series of ultrasonic and mechanical tests and concluded there is a correlation of bond strength with the splitting of the lowest adherend resonance dip and also with the frequency of the highest split resonance (Figure 12).

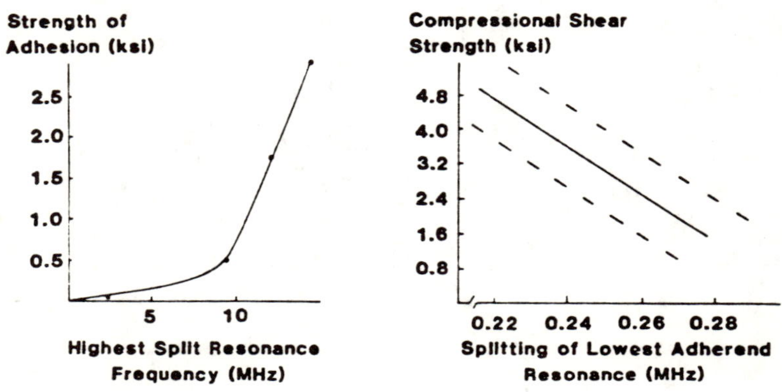


Figure 12. Correlations of Adhesive Bond Strength with Features in the Ultrasonic Reflection Spectrum (after Alers, Elsley and Flynn [17]).

## Surface Roughness

Rough surfaces may be categorized as having random roughness or having a periodic texture. Ultrasonic spectroscopy has been shown to be useful for replacing mechanical profilometer measurements of such surfaces (de Billy, et al [18], Quentin, et al [19] and Jungman, et al [20]).

An ultrasonic transducer is operated in a pulse-echo mode to record the intensity of backscattered waves as a function of angle (Figure 13). These plots (for a fixed frequency) demonstrate the locus of data for a surface having a certain roughness are separate from loci for surface with differing rms roughness.

Experimental measurements to assess the spatial periodicity of a surface are identical to those described for random roughness -- the analysis differs. The surface is assumed to act as a diffraction grating. Peaks in the backscattered signal are found at frequencies

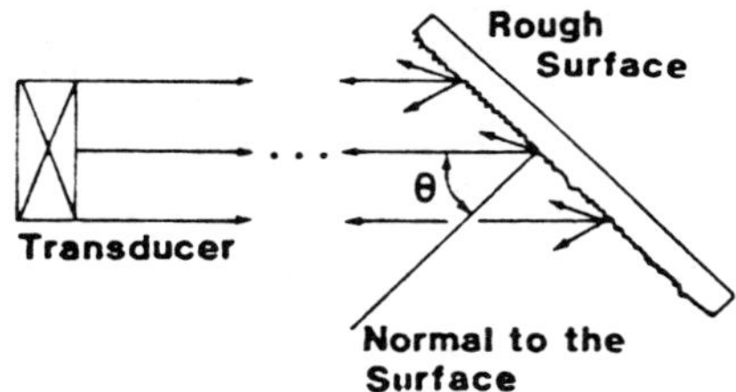


Figure 13. Geometry for Measurement of Backscattered Ultrasound from Rough Surfaces (after de Billy, et al [18]).

$$f_m = m \cdot \frac{c}{2d_1} \cdot \frac{1}{\sin\theta}$$

Since the velocity of the ultrasonic wave (c) in the coupling material is known, as is the order of diffraction (m); the grating spacing ($d_1$) may be found. Imperfect gratings are easily identified by extra peaks in the spectrum at a fixed angle (Figure 14).

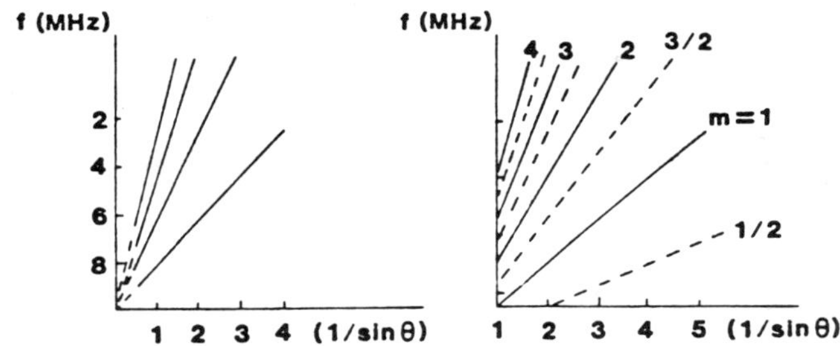


Figure 14. Plots of Frequency of Diffraction Peaks Versus the Reciprocal of the Sine of the Backscattered Angle. (a) perfectly periodic surface; b) a surface having an imperfectly periodic texture (after Quentin, et al [19]).

## Surface Breaking Cracks

An ultrasonic surface (Rayleigh) wave directed toward a surface-breaking defect will interact with the flaw in a manner dependent on the ratio of defect depth to ultrasonic wavelength ($k_R d$). For low $k_R d$ the wave is reflected by the bulk of the defect; whereas, for high $k_R d$ the wave follows the surface of the crack. Partial reflection and transmission occur at each feature of the defect. Measurement of the ensemble of reflected (or transmitted) waves discloses a modulated frequency spectrum. The major frequency of the modulation may be used to nondestructively infer the depth of the flaw. Figure 15 illustrates crack depth measurement via a modified cepstral technique.

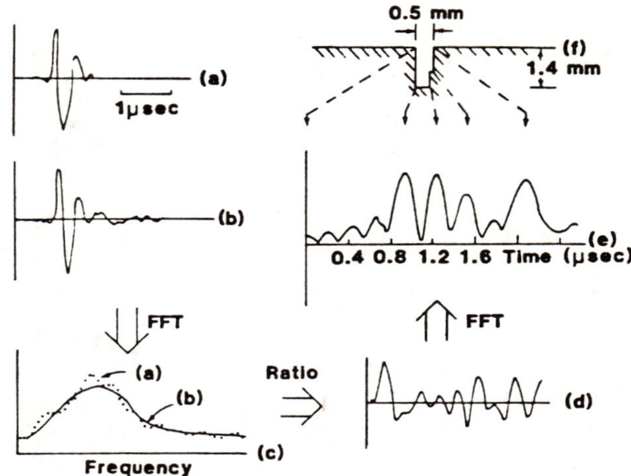


Figure 15. Inference of Crack Depth from Reflected Surface Wave Signals. a) Echo from the crack mouth, b) ensemble of echoes from crack, c) spectra of signals in (a) and (b), d) ratio of spectra in (c) minus unity, e) Fourier transform of (d), peaks correspond to features on the simulated crack (f) (after Hudgell, et al [21]).

Subsurface Material Gradients

Rayleigh waves are not strictly constrained to the surface--they penetrate beneath it to a depth dependent on their wavelength. Since the 'probing' depth depends on frequency, interactions with subsurface defects or gradients will leave an imprint on the spectrum of transmitted frequencies.

Inversion of surface wave dispersion data to find subsurface elastic constant or density gradients have been carried out by Richardson and Tittmann [22] and Szabo [23].

Strength-Related Material Properties

Ultrasonic spectrum analysis techniques have been utilized to determine the strength of composite materials. Stone and Clarke [21] described a method of determining the void content of carbon-fiber reinforced plastic from ultrasonic attenuation measurements. They found a close correlation between attenuation and interlaminar shear strength. Replotting their data as attenuation spectra for various percentage void content (Figure 16) shows ($d\alpha/df$) increases with increasing void content (decreasing strength).

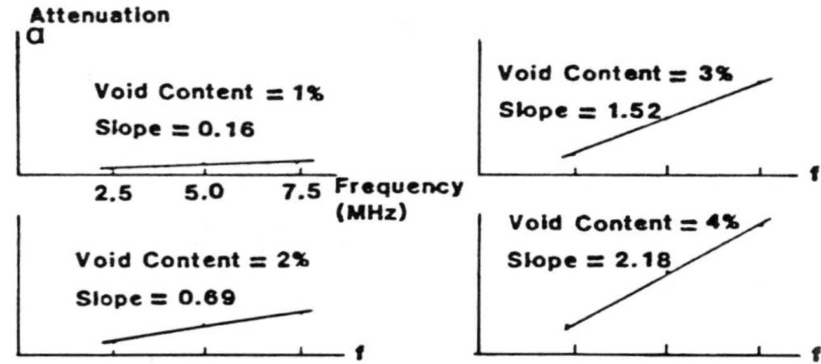


Figure 16. Attenuation Spectra for Composite Panels with Differing Void Content (replotted from data by Stone and Clarke [24]).

## Fracture Mechanics Parameters

Several authors (Richardson [28]; Budiansky and Rice [29]) have described techniques for estimating the fracture mechanics parameter, $k_I$ of a material from long wavelength scattering measurements. A polynomial is fitted to the deconvolved spectrum of the scattered waves. The coefficient of the frequency-squared term can be used to infer the longest dimension of the void, and

$$k_I \times (\pi \text{ longest dimension of the void})^{1/2}$$

## Fracture Toughness

Fracture toughness, a material property, may be expressed as the critical stress intensity ($K_c$) at which a crack will propagate abruptly. It is known that fracture toughness is related to microstructure, and since ultrasonic attenuation ($\alpha$) is also dependent on microstructure, it is reasonable to assume $K_c$ and $\alpha$ are related.

Vary [25] found a linear relationship between the slope of the ultrasonic attenuation coefficient and the plane strain fracture toughness for two types of maraging steel and for a titanium alloy (Figure 17).

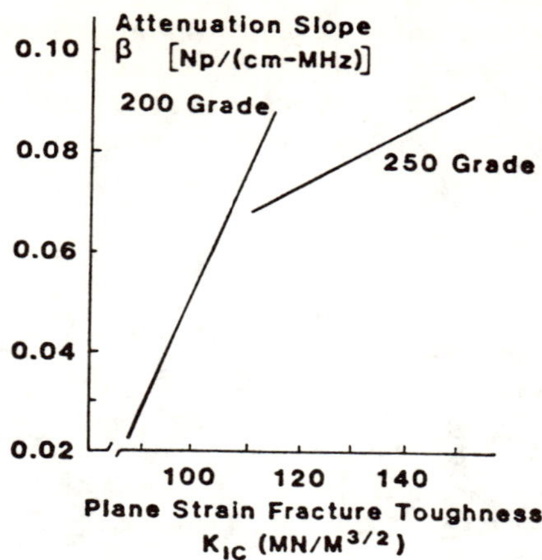


Figure 17. Correlation of Attenuation Slope with Plane Strain Fracture Toughness for Two Maraging Steel Specimens (after Vary [25]).

Microstructure

Because the interactions of elastic waves and a material are wavelength dependent, the microstructure (grain size distribution) is amenable to study by ultrasonic spectroscopic techniques. Attenuation is particularly sensitive to changes in the average grain size ($\overline{D}$) and ultrasonic frequency (f) (Table 1). Entries in the table having coefficients $A_i$ and $B_i$ represent, respectively, scattering and absorption.

Table 1. Functional Dependence of Ultrasonic Attenuation

| Wavelength Range | Functional Dependence |
|---|---|
| $\lambda > 2\pi\overline{D}$ | $B_1f + A_4\overline{V}f^4$ |
| $\lambda < 2\pi\overline{D}$ | $A_2\overline{D}f^2$ |
| $\lambda << D_{min}$ | $B_1f + B_2f^2 + A_o/\overline{D}$ |

$B_1f$ is the elastic hysteresis loss; $B_2f^2$ is the thermoelastic loss, and $\overline{V}$ is the average grain volume.

As noted in the table the attenuation mechanisms depend on grain dimensions as well as frequency. Thus, an ultrasonic wave traversing an attenuating media will bear the imprint of interaction with the grain size distribution. Inference of the average grain size is carried out by carefully determining the attenuation over a wide range of frequencies. The coefficients of polynomials fitted to the attenuation function and the range of frequencies over which the fit is valid may be utilized to characterize the material and to infer the average grain dimension. A good deal of frequency dependent attenuation data is available; however, there is a scarcity of investigations carrying the process through to determining grain size.

Quantitative grain scattering formulae have been proposed and investigated [26,27]. These expressions involve the elastic constants of the individual grain, the density, frequency and grain size. One theory, proposed by Roney [27], may be applicable through all three ranges of the ratio of wavelength to grain dimension.

## CONCLUSION

The sole purpose of an ultrasonic spectroscopic system is to determine the properties of the material or system being tested. Implicit in the assumption that useful information can be derived from the output of a spectroscope is the relevant properties of the material are, or can be, related to parameters which are frequency dependent. Many of the applications have been reviewed in this paper. Each use of ultrasonic spectroscopy has arisen because a definite relation (theoretical or experimental) has been found between the frequency response and an important material characteristic.

## REFERENCES


1. Gericke, O. R. "Spectrum and Contour Analysis of Ultrasonic Pulses for Improved Nondestructive Testing," Technical Report WAL TR 830.5/1 (Watertown Arsenal Laboratories, 1960).

2. Fitting, D. W. and Adler, L. *Ultrasonic Spectral Analysis for Nondestructive Evaluation* (New York: Plenum Publishing Co., 1981).

3. Frederick, J. R. and Seydel, J. A. "Improved Discontinuity Detection Using Computer-Aided Ultrasonic Pulse-Echo Technique," *Bulletin Number 185 of the Welding Research Council* (1973).

4. Rhyne, T. L. "Radiation Coupling of a Disk to a Plane and Back or a Disk to Disk: An Exact Solution," *J. Acoust.*

*Soc. Am.* 61(2):318-324 (1977).

5. Tittmann, B. R. "Ultrasonic Measurements for the Prediction of Mechanical Strength," *NDT International* 17-22 (1978).

6. Adler, L. and Lewis, K. "Models for the Frequency Dependence of Ultrasonic Scattering from Real Flaws," *Proc. ARPA/AFML Rev. Prog. Quant. NDE,* AFML-TR-77-44, 180-186 (1977).

7. Achenbach, J. D., Adler, L., Lewis, D. K. and McMaken, H. "Diffraction of Ultrasonic Waves by Penny-Shaped Cracks in Metals: Theory and Experiment," *J. Acoust. Soc. Am.* 66:1848 (1979).

8. Adler, L. and Achenbach, J. D. "Elastic Wave Diffraction by Elliptical Cracks: Theory and Experiment," *J. Nondes. Eval.* 1(2):87-99 (1980).

9. Pao, Y. H. and Sachse, W. "Interpretation of Time Records and Power Spectra of Scattered Ultrasonic Pulses in Solids," *J. Acoust. Soc. Am.* 54:1478-1486 (1974).

10. Sachse, W. "The Scattering of Elastic Pulses and the Nondestructive Evaluation of Materials," *Mat. Eval.* 106: 83-89 (1977).

11. Lloyd, E. A. "Nondestructive Testing of Bonded Joints," *Nondestructive Testing (London),* 331-334 (1974).

12. Chang, F. H., Couchman, J. C. and Yee, B. G. W. "Transmission Frequency Spectra of Ultrasonic Waves through Multi-Layer Media," *Proc. 1973 IEEE Ultrasonic Symp.* (1973).

13. Scott, W. R. and Gordon, P. F. "Ultrasonic Spectrum Analysis for Nondestructive Testing of Layered Composite Materials," *J. Acoust. Soc. Am.* 62(1):108-116 (1977).

14. Rose, J. L. and Meyer, P. A. "Ultrasonic Procedures for the Determination of Bond Strength," (NTIS AD-A015 291) AFOSR-73-2480A Technical Report for the Air Force Office of Scientific Research (1975).

15. Couchman, J. C., Chang, F. H., Yee, B. G. and Bell, J. R. "Resonance Splitting in Ultrasonic Spectroscopy," *IEEE Trans. Son. Ultrason., SU-25* 5 (1978).

16. Chang, F. H., Flynn, P. L., Gordon, D. E. and Bell, J. R. "Principles and Application of Ultrasonic Spectroscopy in NDE of Adhesive Bonds," *IEEE Trans. Son. Ultrason. SU-23* 5:334-338 (1976).

17. Alers, G. A., Elsley, R. K. and Flynn, P. L. "Measurement of Strength of Adhesive Bonds," *Proc. ARPA/AFML Rev. Prog. Quant. NDE,* AFML-TR-78-55, 365-370 (1978).

18. deBilly, M., Doucet, J. and Quentin, G. "Angular Dependence of the Backscattered Intensity of Acoustic Waves from Rough Surfaces," *Ultrasonics International 1975, Conf. Proc*; IPC Science & Technology Press, London (1975).

19. Quentin, G., deBilly, M., Cohen-Tenoudji, F., Doucet, J. and Jungman, A. "Experimental Results on the Scattering of Ultrasound by Randomly and Periodically Rough Surfaces in the Frequency Range 2 to 25 MHZ," *1975 IEEE Ultrasonics Symp., Proc.,* 102-106 (1975).

20. Jungman, A., Cohen-Tenoudji, F. and Quentin, G. "Diffraction Experiments in Ultrasonic Spectroscopy; Preliminary Results on the Characterization of Periodic or Quasi-Periodic Surfaces," *Ultrasonic International 1977 Conf. Proc.*; IPC Science and Technology Press (1977).

21. Hudgell, R. J., Morgan, L. L. and Lamb, R. F. "Nondestructive Measurement of the Depth of Surface-Breaking Cracks Using Ultrasonic Rayleigh Waves," *Br. J. NDT* 16:144-149 (1974).

22. Richardson, J. M. and Tittmann, B. R. "Deducing Subsurface Property Gradients from Surface Wave Dispersion Data," *Proc. 1975 IEEE Ultrason. Symp.,* 488-491 (1975).

23. Szabo, T. L. "Obtaining Subsurface Profiles from Subsurface-Acoustic-Wave Velocity Dispersion," Air Force Cambridge Res. Lab. Report AFCRL-TR-75-0273 (NTIS #AD-A010 081/8ST) and *J. Appl. Phys.* 46(4):1448-1454 (1975).

24. Stone, D. E. W., Clarke, B. "Ultrasonic Attenuation as a Measure of Void Content in Carbon-Fibre Reinforced Plastics," *Non-Destr. Test. (London)* 8(3):137-145 (1975).

25. Vary, A. "Feasibility of Ranking Fracture Toughness by Ultrasonic Measurements," *Proc. 1975 IEEE Ultrason. Symp.* 558-590 (1975).

26. Papadakis, E. P. "Ultrasonic Attenuation Caused by Scattering in Polycrystalline Media," Chap. 15 of *Physical Acoustics*, Vol. IV, Part B, W. P. Mason, Ed., (New York: Academic Press, 1968), p. 269-329.

27. Serabian, S. and Williams, R. S. "Experimental Determination of Ultrasonic Attenuating Characteristics Using the Roney Generalized Theory," *Mat. Eval.* 55-62 (1978).

28. Richardson, J. M. "Direct and Inverse Problems Pertaining to the Scattering of Elastic Waves in the Rayleigh (Long Wavelength) Regime," *Proc. ARPA/AFML Rev. of Prog. in Quant. NDE,* AFML-TR-78-205, 332-340 (1978).

29. Budiansky, B. and Rice, J. R. "On the Estimation of a Crack Fracture Parameter by Long-Wavelength Scattering," *J. Appl. Mech.* 45(2) (1978).

CHAPTER 11

# PHOTOELASTIC VISUALIZATION IN THE DEVELOPMENT OF QUANTITATIVE ULTRASONICS

*C. P. Burger and A. J. Testa*
Department of Engineering Science and Mechanics
and Engineering Research Institute
Iowa State University
Ames, Iowa 50011

*A. Singh*
Southwest Research Institute
San Antonio, Texas 78284

Dynamic photoelasticity gives full field visualization of the stresses produced when an elastic wave travels through a transparent birefringent solic. A sequence of high speed photographs from a Cranz-Schardin camera can then be used to observe the interactions between elastic waves and different kinds and shapes of defects. Spectroscopic techniques can be applied directly to the optical data to extract information on the "transfer function" for a particular defect, but the resolution of contemporary systems is still generally poor. By combining the physical information from the simulated optical observations with signals from ultrasonic tests on prototype materials, however, much information is gained towards explaining wave/flaw interactions. As a result, improved pattern recognition and signal processing procedures can be developed that will lead to new and improved capabilities in quantitative NDE.

The paper describes the system that was used for dynamic photoelastic visualization of stress waves and explains how optical spectroscopy was applied to the data. Examples are given for the interactions between Rayleigh waves and surface slots. The optical data is then used to develop a signature analysis procedure in the frequency domain for finding the depths of shallow slots.

Developments in quantitative ultrasonics depends on a thorough understanding of the interactions between elastic waves and defects. This requires progressive information of the dynamic stress field so that the limited information contained in ultrasonic observations of the surface effects of the waves can be extrapolated backwards to identify the changes wrought on the waves by the defects. The significant features of the surface signals that best characterize a defect, with respect to its structurally important features, can then be identified and used to develop new and effective ultrasonic evaluation procedures.

Despite the significant advances during recent years in analytical and numerical procedures, their value in practical engineering terms are still small. The inverse problem, for example, has barely been solved for the most elementary of defects. There is a clear need for new ways of acquiring the knowledge and insights that are necessary if ultrasonics is to fulfill its promise as the ultimate tool for NDE. Photoelastic visualization interactively coupled with ultrasonic experimentation and with analytical support, is one such way. This paper describes one method of visualization and illustrates how the insights so gained led to a promising new approach for characterizing shallow surface cracks.

## DYNAMIC PHOTOELASTICITY

Dynamic photoelasticity provides full-field two-dimensional photographs of the elastic waves in sheets of birefringent material. For the visualization of acoustic waves, the preferred model material should have acoustic properties similar to those of the prototype materials. Two suitable materials are quartz and glass. Unfortunately, their birefringence is so low, their wave speeds so high and their wavelengths so short, that they can at present provide only qualitative information [1]. The more traditional polymeric model materials have low moduli so that excitation with ultrasonic transducers produce stress levels that are too low to produce useful levels of birefringence. However, when the impulse source causes large enough displacements, a sufficient number of photoelastic fringes can be produced to resolve the stress field and allow for effective visualization of the waves. Wave speeds in these materials are also lower and the wavelengths longer, so that exposure times of the order of 1 μs can be used to effectively freeze the wave.

The most widely used impulse source is a small explosion. The model is usually damaged slightly in the vicinity of the explosion and needs repair or replacement before a second shot can be fired. Stroboscopic methods, which rely on repetitive impulses and flash exposures with different delay times after the impulse, are, therefore, impractical and prohibitively expensive. They also do not offer sufficiently accurate reproducibility when explosive sources are being used. The Cranz-Schardin type multiple spark camera [2,3] with 16 consecutive high voltage spark gaps can take 16 sequential pictures of the progressive stress field from a single model with one explosion. The individual exposure times for each dynamic picture is 600 ns and the elapsed time between consecutive photos can be adjusted from 1 μs to 14 μs. Thus, complete information on the interactions of an elastic wave with a defect can be obtained from a small number of design models.

The photoelastic models used in the investigation reported here were made from 6.3 mm (1/4 in.) thick sheets of a polyester material "Homalite 100" (s.g.L. Homalite Corp.) with properties as shown in Table 1. The nature and dimensions of the models are shown in Figure 1. They were chosen to avoid the mixing of stray reflections of dilitational (P) or distortional (S) waves from the edges of the plate with the waves being studied. The explosive was lead azide in charges of 45 mg set off 175 mm from the defect with which the wave will interact. This is far enough to allow the surface (R) wave to separate from the shear wave before they reach the defect. The charge weight is chosen so that it just carries a small amount of fracture to occur at the site of the explosion. Excess energy over and above the amount needed to initiate fracture is absorbed in the fracture process. Beyond the fracture zone an initial plastic wave will propagate outward. This wave has a well-defined magnitude which is not sensitive to small changes in the intensity of the explosion. It soon attenuates to an elastic wave of reproducible shape and magnitude [3]. This is the wave which is observed in all the photographs reproduced in this paper.

Table 1. Dynamic Properties of Homalite-100

| | |
|---|---|
| Velocities in Homalite-100 Plate | m/s (in./s) |
| Longitudinal Wave | $2.1 \times 10^3$ (83,000) |
| Shear Wave | $1.22 \times 10^3$ (48,000) |
| Rayleigh Wave | $1.11 \times 10^3$ (43,500) |
| Elastic Modulus (Tensile) | $2.4 \times 10^9$ N/m$^2$ ($3.5 \times 10^5$ psi) |
| Elastic Modulus (Compressive) | $4.5 \times 10^9$ N/m$^2$ ($6.5 \times 10^5$ psi) |
| Specific Gravity | 1.23 |

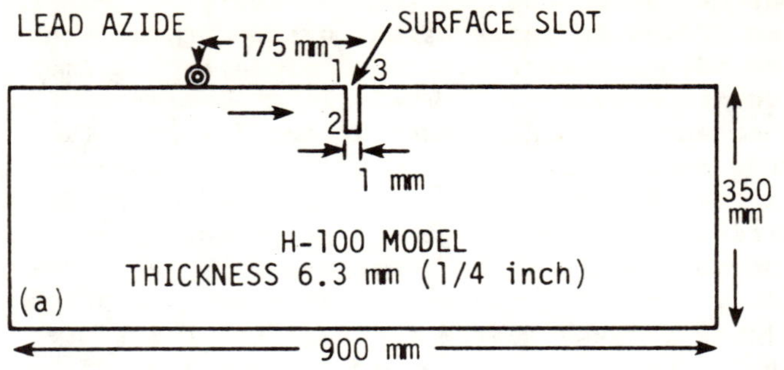


Figure 1. Dimension of photoelastic models.

A typical sequence of dynamic photographs is shown in Figure 2. Only the last 12 of the original 16 frames are reproduced. The interval between the pictures is 11.5 μs, and the distance between the vertical grid lines is 20 mm.

In frames 5 to 8 of Figure 2 the incident Rayleigh (R) wave approaches a 2.8 mm slot (1 mm wide). The leading dilitational (P) on Von Smidt (PS) waves are present as weak lines of 1/2 fringe ahead of the R-wave. They can be seen best in frames 6 and 7. The main shear wave (S) in the body of the plate is clearly identifiable as the circular lines below the R-wave. The PS wave is tangent to S.

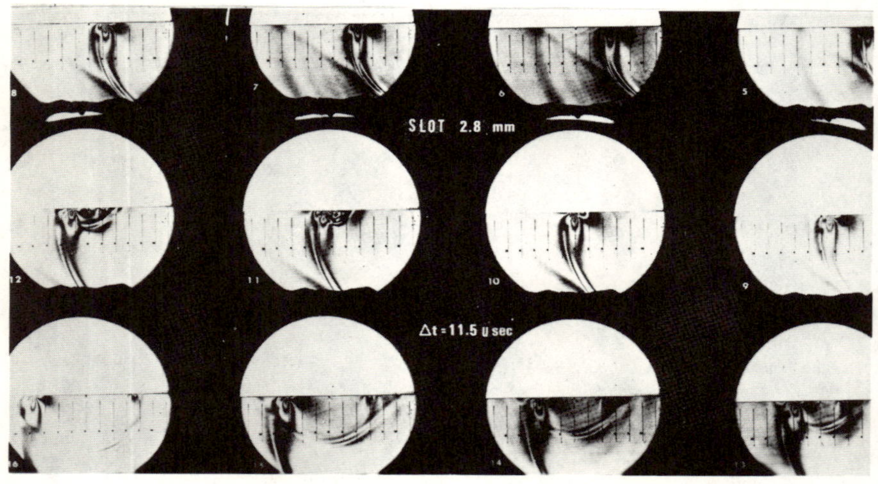


Figure 2. Typical sequence of frames from spark camera [16].

The characteristics of the R-wave is best obtained from frame number 8. Along the top edge of the plate the tangential surface stress is proportional to the fringe order N(5). Thus, if the fringe order is plotted along this surface, a spatial image is obtained of the incident R-wave at the boundary. A Fast Fourier transformation of this wave will give its frequency spectrum [4]. For the tests reported in this paper, the R-waves were broadband pulses with significant frequencies from about 14 kHz to 100 kHz and peak frequencies near 25 kHz [4,5]. With the velocities in Table 1, the peak frequency corresponds to a wavelength of 44.4 mm. Below the surface the fringe order $N \alpha (\sigma_1 - \sigma_2)$; i.e., N is proportional to the in-plane maximum shearing stress. It is clear from the photographs that the fringe orders associated with the R-wave attenuate rapidly with depth below the surface and is less than 1/2 at depths greater than about 60 mm or approximately 1.4 x peak wavelength. The real depth is hard to identify because the lower portion of the Rayleigh wave is still coupled to the shear wave in this region.

Figure 3 is an enlargement of an incident Rayleigh wave approaching a 9.9 mm deep slit. It shows the characteristic pattern of the surface wave.

The next four frames of Figure 2(Nos. 9 through 12) show the wave interacting with the slot. In the last four frames

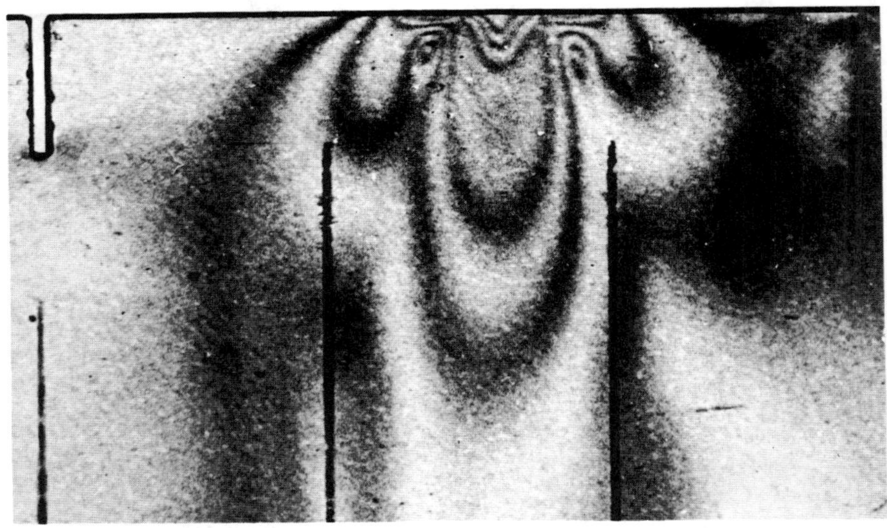

Figure 3. Incident Rayleigh wave and 9.9 mm deep slot.

(Nos. 13 through 16) the transmitted, mode converted and reflected waves separate and travel away from the slot. The dominant characteristic is the strong mode converted shear wave which travels downward in the 5 o'clock direction from the slot tip.

Four selected frames from the sequences for three different slot depths are shown in Figure 4. Each of the four rows show the waves in approximately the same position with respect to the different slots. The differences in the effect that the different slots have on the waves are clearly illustrated. In the top row the incident R-wave had just reached the slot. The deeper portion of the wave will be cut off from the near surface portion. In the second row the deeper, low frequency portion of the wave has almost finished its interaction with the tip of the slot. The two outer lobes of the incident wave are at either side of the tip and the main lobe is very nearly at the tip. In the third row the transmitted Rayleigh wave is just established at the far surface and the characteristic fringe pattern can be readily identified for the 2.8 mm deep slot. In the last row the three characteristic waves that exist after the interaction can be seen. They are the transmitted R-wave along the far edge, the mode converted shear wave traveling downward into the plate from the slot and the reflected R-wave. There are also many secondary waves present.

The shape of the transmitted waves at the top edge for the approximate times shown in the fourth row of Figure 4 were plotted and then frequency spectra computed. The results of this "optical spectroscopy" procedure and their interpretation was reported in References 4 and 5. It established the postulate that a slot will act as a low pass filter for a broadband R-wave with depth greater than the slot. The deeper portions of the R-wave will appear at the far surface ahead of the main body of the R-wave that was cut off by the slot. Thus, after the waves have separated from the slot, the transmitted wave will consist of two R-waves traveling together, the one slightly behind the other. The first wave will not contain the high frequency components of the incident wave which was cut off by the slot, while the trailing R-wave will have all the frequency components but with reduced energy in the lower frequencies because some of that energy was in the deeper portion of the wave and is, therefore, in the leading R-wave. The cut-off frequency of the leading R-wave; i.e., the frequency at which the wave is almost devoid of higher frequency components, will relate to the depth of the slot. This is shown as points $C_1$, $C_2$, $C_3$ and $C_4$ of Figure 5 [6].

This postulate was confirmed by running ultrasonic tests on slots in steel [6,7]. Testa extended these tests to much

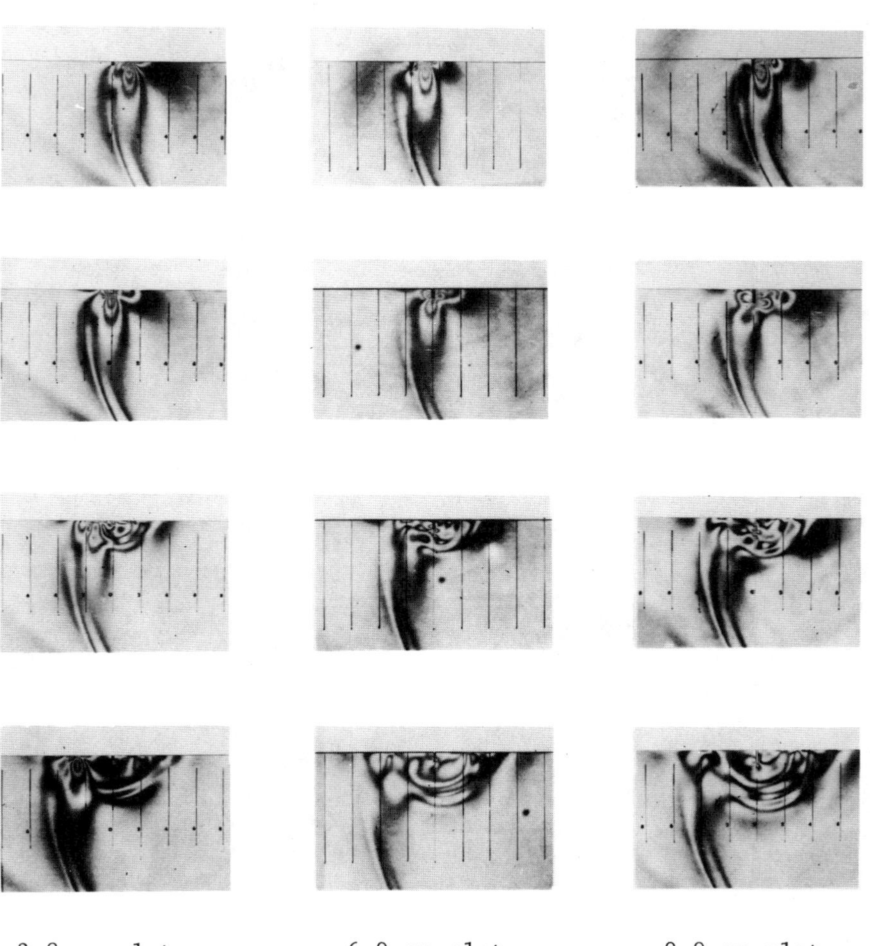


Figure 4. Selected frames from three models.

narrower slots made by Electro Discharge Machining (EDM) and with an open fatigue crack [8].

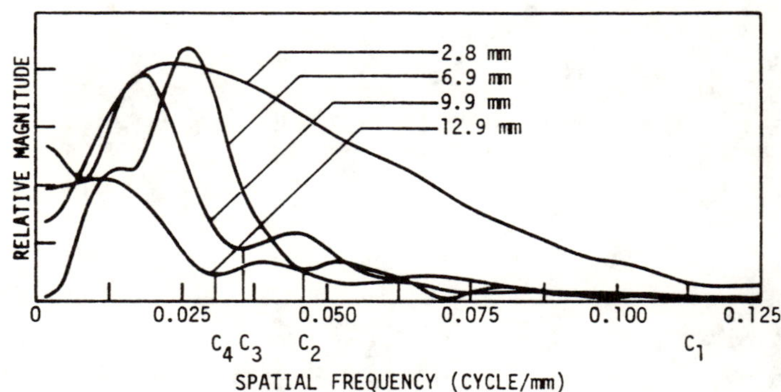


Figure 5. Frequency spectra for the leading transmitted R-wave from surface fringe orders on four slots [6].

Figure 6 is a selection of four frames from three models that further illustrate the dynamic photoelastic technique.

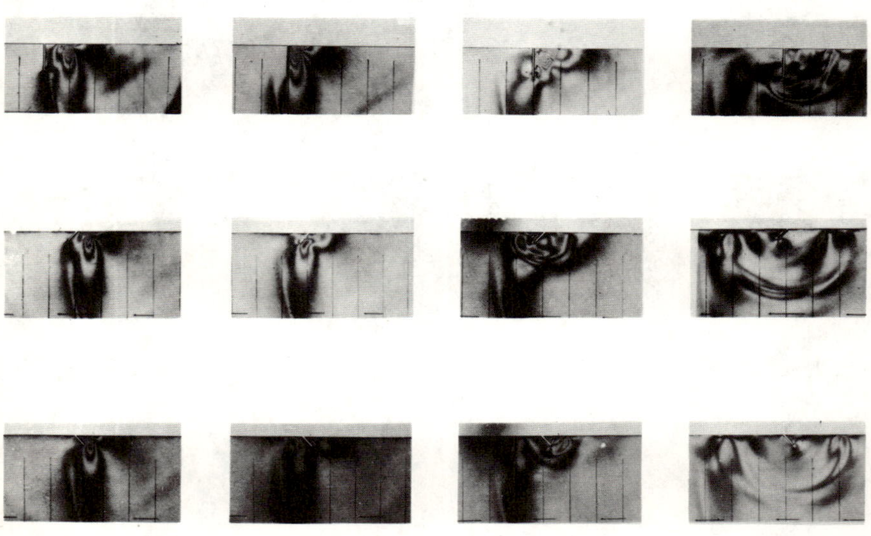

Figure 6. Dynamic fringe patterns for a 22.8 mm deep slot and for two slanting slots 10 mm long and 135° and 45°.

For the deep slot the transmitted surface wave is very weak, as expected. For the two slanted slots the relative amplitudes of the reflected on transmitted waves varies with angle, but the frequency spectrum of the leading transmitted wave had the correct cut-off frequency for a slot 10 gm $45^{o}$ = 7.1 mm deep.

## ACKNOWLEDGMENTS

This research was supported by the U.S. Department of Energy, Contract Number W7405 ENG82, the Office of Basic Energy Sciences, Division of Material Sciences (AK 01 02), and by the Engineering Research Institute and Department of Engineering Science and Mechanics at Iowa State University.

## REFERENCES

1. Hall, K. G. "Crack Depth Measurements in Rail Steel by Rayleigh Visualization," *Non-Destructive Testing* 9(3): 121-126 (1976).

2. Riley, W. F. and Dally, J. W. "Recording Dynamic Fringe Patterns with a Cranz-Schardin Camera," *Experimental Mechanics* 9:27N-33N (1969).

3. Ligon, J. B. *A Photomechanics Study of Wave Propagation* (Iowa: Ph.D. Dissertation, Iowa State University, 1971).

4. Singh, A., Burger, C. P., Schmerr, L. W. and Zachary, L. W. "Dynamic Photoelasticity as an Aid to Sizing Surface Cracks by Frequency Analysis," *Mechanics of Non-destructive Testing*, W. W. Stinchcomb, Ed. (New York: Plenum Publishing Corp., 1980).

5. Burger, C. P., Testa, A. and Singh, A. "Dynamic Photoelasticity as an Aid in Developing New Ultrasonic Test Methods," (Dearborn, Michigan: Proc. Soc. for Experimental Stress Analysis, Spring Meeting, June 1981).

6. Singh, A. *Crack Depth Determination by Ultrasonic Frequency Analysis Aided by Dynamic Photoelasticity* (Iowa: M.S. Thesis, Iowa State University, 1980).

7. Burger, C. P. and Singh, A. "An Ultrasonic Technique for Sizing Surface Cracks," (San Diego, California: Proc. DARPA/AF Review of Progress in Quantitative NDE, July 1980).

8. Burger, C. P. and Testa, A. "Rayleigh Wave Spectroscopy to Measure the Depth of Surface Cracks," (San Antonio, Texas: Proc. 13th Symposium on Non-destructive Evaluation, April 1981).

INDEX